Control Engineering

Second edition

TUTORIAL GUIDES IN ELECTRONIC ENGINEERING

Series editors
Professor G.G. Bloodworth, *University of York*
Professor A.P. Dorey, *University of Lancaster*
Professor J.K. Fidler, *University of York*

This series is aimed at first- and second-year undergraduate courses. Each text is complete in itself, although linked with others in the series. Where possible, the trend towards a 'systems' approach is acknowledged, but classical fundamental areas of study have not been excluded. Worked examples feature prominently and indicate, where appropriate, a number of approaches to the same problem.

A format providing marginal notes has been adopted to allow the authors to include ideas and material to support the main text. These notes include references to standard mainstream texts and commentary on the applicability of solution methods, aimed particularly at covering points normally found difficult. Graded problems are provided at the end of each chapter, with answers at the end of the book.

1. Transistor Circuit Techniques: discrete and integrated (3rd edition) — G.J. Ritchie
2. Feedback Circuits and Op Amps (2nd edition) - D.H. Horrocks 3. Pascal for Electronic Engineers (2nd edition) — J. Attikiouzel
4. Computers and Microprocessors: components and systems (3rd edition) — A.C. Downton
5. Telecommunication Principles (2nd edition) — J.J. O'Reilly
6. Digital Logic Techniques: principles and practice (2nd edition) — T.J. Stonham
7. Instrumentation: Transducers and Interfacing — B.R. Bannister and D.G. Whitehead
8. Signals and Systems: models and behaviour (2nd edition) — M.L. Meade and C.R. Dillon
9. Power Electronics (2nd edition) — D.A. Bradley
10. Semiconductor Devices (2nd edition) — J.J. Sparkes
11. Electronic Components and Technology (2nd edition) — S.J. Sangwine
12. Control Engineering (2nd edition) — C. C. Bissell
13. Basic Mathematics for Electronic Engineers: models and applications — J.E. Szymanski
14. Software Engineering — D. Ince
15. Integrated Circuit Design Technology — M.J. Morant

Control Engineering

Second edition

C. C. Bissell
Lecturer in Electronics
Faculty of Technology
The Open University
UK

CRC Press
Taylor & Francis Group
Boca Raton   London   New York

CRC Press is an imprint of the
Taylor & Francis Group, an informa business

Reprinted 2010 by CRC Press
CRC Press
6000 Broken Sound Parkway, NW
Suite 300, Boca Raton, FL 33487
270 Madison Avenue
New York, NY 10016
2 Park Square, Milton Park
Abingdon, Oxon OX14 4RN, UK

Catalog record is available from the Library of Congress

Visit the CRC Press Web site at www.crcpress.com

No claim to original U.S. Government works
International Standard Book Number 0-412-57710-0
Library of Congress Card Number 94-70238

Contents

Preface to the second edition

Apart from a few minor corrections and modifications, Chapters 1–7 and the first half of Chapter 8 remain unchanged from the first edition. Chapter 8 has been extended to include a detailed discussion of disturbance rejection and cascade control, and four completely new chapters have been added on the modelling and design of discrete control systems. I am grateful to the Open University for permission to base much of Chapters 9 and 10 on Block 1 of course PMT604 Real-Time Control. Thanks also go to my colleagues Roger Loxton (for the cascade control example of Chapter 8) and Chris Dillon (for some of the material in Chapters 9 and 10, and for reading and commenting on the second edition in draft form).

Preface

The appearance of this book in the series 'Tutorial Guides in Electronic Engineering' is a reflection of the importance attached to control in electronics and electrical engineering curricula. Yet control engineering is essentially interdisciplinary in nature, and plays a fundamental role in many other areas of technology. I have therefore tried to make this text equally relevant to readers whose main interest lie outside electronics, by concentrating on general systems characteristics rather than on specific implementations.

I have restricted myself to the 'classical' approach to single-input, single-output systems, since I feel this is the most appropriate subject matter for a first course in control. However, the Tutorial Guide style, with its detailed treatment of simple design examples, should also render the text useful to practising engineers who need to revise and apply dimly remembered material – or even to those whose training did not include control.

The reader is assumed to be familiar with complex numbers, phasors, and elementary calculus. Apart from these topics, the mathematical requirements are few, although prior knowledge of simple first- and second-order linear differential equations would be useful.

Where possible I have tried to indicate how computer-based tools can reduce the labour involved in control system design, although limitations of space have precluded detailed description. However, CAD software or other computer-based approaches can only be as effective as the understanding and skill of the user. In the chapters dealing with aspects of design I have tried to develop such understanding by dealing with a limited number of examples in depth, rather than giving a cursory treatment of a wider range of material. Nevertheless, the examples have been chosen to illustrate most of the major classical techniques of feedback control, including the distinctive features of digital implementations.

My approach has been strongly influenced by the Open University course T391 Control Engineering and its successor T394, and it is a pleasure to record my debt to other members of those course teams. I am particularly grateful for many hours of discussion with Chris Dillon, who has read and commented on draft chapters with great perception, and whose ideas have contributed substantially to the final version. Thanks are also due to series editor Professor Kel Fidler for support and guidance.

Acknowledgements

Figures 3.21, 3.22, 4.21, 4.22, 8.14, 8.16, 8.18–8.22, 9.3, and 10.1–10.8 are reproduced by kind permission of the Open University (© Open University Press 1984, 1984, 1978, 1978, 1978, 1978, 1978, 1978 and 1987, respectively). The CODAS-II package is produced by Golten & Verwer Partners (33 Moseley Road, Cheadle Hulme, Cheshire, SK8 5HJ, UK); Matlab and Simulink are trademarks of The MathWorks, Inc (Cochituate Place, 24 Prime Park Way, Natick, Mass. 01760 USA); Macsyma is copyright Macsyma Inc/MIT.

Systems, objectives and strategies

1

☐ To introduce the concept of control in an engineering context, and to indicate the wide variety of control tasks in a large engineering system.

☐ To describe the three common control strategies – open loop, feedforward, and feedback (or closed-loop control).

Objectives

Introduction

The word 'control' is used in many different contexts. We talk of quality control, financial control, command and control, production control, and so on – terms which cover an enormous range of activities. Yet all these types of control, if they are to be successful, have certain features in common. One is that they all presuppose the existence of a *system* whose behaviour we wish to influence, and the freedom to take actions which will force it to behave in some desirable way. For example, for the manager of a large chemical plant the system of interest may be the entire plant, as illustrated in Fig. 1.1. The *inputs* to the system, which we assume the manager can influence, are the various flows of energy and raw materials into the plant; the *outputs* are not only the finished products but also the waste, environmental effects, and so on. Note that there are also *disturbance inputs*, which the manager cannot control, such as market fluctuations, changes in the environment, etc., and these will also affect the plant outputs. For another engineer in the same plant, however, the system of interest might be one particular reaction vessel and specifically, the design of a control system to maintain

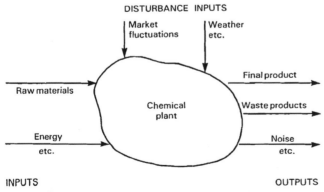

Fig. 1.1 A chemical plant considered as a single system.

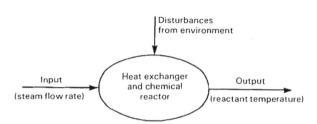

Fig. 1.2 A reactor subsystem of the chemical plant.

In general, a system can be defined loosely as the set of interconnected elements which are of interest for some specific purpose.

the reactants at a constant temperature by adjusting the flow of steam to a heater, as represented in Fig. 1.2. From a control point of view there is a single input (steam flow rate) and a single output (reactor temperature); in addition, there will again be disturbances caused by various outside factors such as unwanted fluctuations in the steam supply or changes in the ambient temperature.

No control system can be designed without a clear specification of control *objectives*. For a chemical plant as a whole, the ultimate control objective might be to produce a final product meeting quality specifications, while minimizing costs. For the temperature control system the objective might be to remain within a certain temperature range under specified operating conditions. In each case, however, there will be limitations or *constraints* on what can be achieved; not only limits to the physical capabilities of the equipment being used, but also, for example, economic, legal, and safety constraints.

Control engineering can perhaps be summed up as the design and implementation of automatic control systems to achieve specified objectives under given constraints. For a complex system, the overall objectives and constraints will need to be translated into performance specifications for the various subsystems – ultimately into control system specifications for low-level subsystems, such as individual chemical reactors in the chemical plant example.

Control engineering as a discipline is characterized by a common approach to a great variety of control tasks, and by a set of mathematical tools which have proved to be generally applicable. Computers are used widely to implement control schemes, and an increasing knowledge of information technology and software engineering is therefore being demanded of control engineers. Nevertheless, the fundamental requirement is still a thorough understanding of the dynamics of individual control systems. This book will concentrate on those strategies, models and techniques vital to this understanding.

Control strategies

This text will deal with single-input, single-output systems only. Such systems can be represented by a block diagram such as Fig. 1.3, which shows a single-input, single-output process to be controlled.

The term 'process' is used generally to mean any system to be controlled. The term 'plant' is often also used with exactly the same meaning.

In the introduction to this chapter the terms input and output were used very generally to signify any flow of information, energy or material into or out of a system. From now on, however, the terms will be used more precisely. In Fig. 1.3, for example, $u(t)$, the 'input' to the process, represents the variable which is

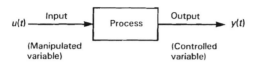

Fig. 1.3 A single-input, single-output process.

adjusted in order to bring about the control action: it is often known as the *manipulated variable*. Similarly $y(t)$, the 'output' of the process, is the variable which the engineer wishes to control in order to fulfil the desired objectives: not surprisingly, it is referred to as the *controlled variable*. Hence in the temperature control system mentioned above the 'input' (manipulated variable) was the rate of flow of steam and the 'output' (controlled variable) was the temperature – even though to a chemical engineer the various reactants and products might be perceived as the system inputs and outputs! To a control engineer, inputs and outputs are defined so as to represent a *signal flow* through the control system, a concept which will become clearer in subsequent chapters.

Returning to Fig. 1.3, then, we can express the goal of all controllers – whether automatic systems or human operators – as attempting to achieve a desired output behaviour $y(t)$ by applying an appropriate control action $u(t)$ to the process. Automatic controllers or control systems do this by using information about the process and the particular operating conditions to determine an appropriate $u(t)$ for a given situation, as represented by Fig. 1.4. Such information might include externally supplied data about operating conditions – such as the desired and current values of $y(t)$, the rate at which $y(t)$ is changing, and so on – but also 'built-in' information which takes into account how the process is likely to behave in response to a particular control action.

The latter, 'built-in' information is derived from a *model* of the process which can be used to predict the variation of $y(t)$ for a given applied $u(t)$. Models of process behaviour are vital if a control system is to be designed which will automatically generate an appropriate control action, and the sort of models commonly used by control engineers will be described in some detail in Chapter 3. First, however, let us examine a number of general approaches or *control strategies* which can be adopted. All require the process to be modelled, but the general characteristics of each strategy can be described without making any particular assumptions about the type of model employed.

The first strategy, known as *open-loop control*, is illustrated in Fig. 1.5. The

Modern automatic control systems are based on the use of digital computers or microprocessors to generate the required control action. Digital and computer control is discussed in detail in Chapters 9–12.

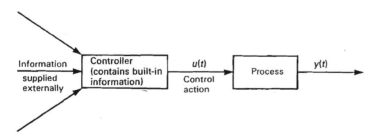

Fig. 1.4 The general control problem.

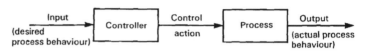

Fig. 1.5 The open-loop strategy.

The precise form of the model used will depend on many factors, including the objectives of the control system. The modelling process will be discussed in detail in Chapter 3.

controller can be thought of as using an 'inverse' model of the process, together with externally-supplied information about the desired output, to determine control action. A simple example should make this clearer. Figure 1.6 shows an open-loop motor speed control system. Suppose that we have a model relating the motor armature voltage to the resulting motor speed. Using such a speed/voltage relationship, we can attempt to design an open-loop controller such that the armature voltage generated in response to a given desired speed (as defined by the position of a control knob, for example) is just what is required, in theory, to produce that particular motor speed. If the motor speed/voltage relationship is modelled by a constant – say $G\,\mathrm{rad\,s^{-1}\,V^{-1}}$ – then the effect of the controller and power amplifier combined must be to produce $1/G$ volts for each $\mathrm{rad\,s^{-1}}$ of demanded speed. In this simple case the open-loop controller attempts to implement the exact inverse model, $1/G$.

There are various drawbacks to this type of control strategy, however. If the load on the motor changes, the speed will alter even if the demanded speed, and hence the armature voltage, is held constant. Furthermore, the characteristics of the motor may vary with time – for example, the speed obtained for a given voltage when the motor is cold may be very different from that when the lubricating oil in the bearings has reached its normal operating temperature. Open-loop control cannot compensate for either *disturbances* to the system (such as a varying load) or changes in plant parameters (such as varying friction in the bearings).

One way of compensating for disturbances is to measure them and make corresponding changes to the control action, as illustrated in general terms in Fig. 1.7. Here one input to the *controller* represents the desired behaviour of the process in some way. The control action taken by the controller is determined not only by using a model of how the process behaves, but also by taking into account the measured disturbances. In the case of the temperature control system of Fig. 1.2, for example, disturbances to the steam supply system might conceivably be measured and the value used to open or close a supply valve as appropriate, again using a model of the reaction vessel to determine the compensating control

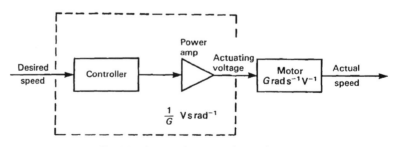

Fig. 1.6 An open-loop speed control system.

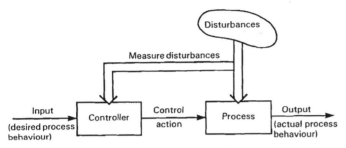

Fig. 1.7 The feedforward strategy.

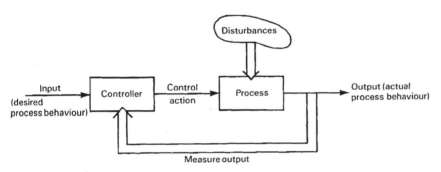

Fig. 1.8 The closed-loop (feedback) control strategy.

action. This type of control attempts to compensate for disturbances before they have any effect on the system output, and is known as *feedforward control*. Feedforward control can be a very effective strategy if the disturbances have a known effect and can be easily measured. If there are too many disturbances, however, or they cannot easily be measured, then feedforward control is not effective. Furthermore, feedforward control cannot compensate for any changes in the plant characteristics which cannot be measured and treated as a disturbance.

The most common control stategy is *feedback* or *closed-loop control*, illustrated in Fig. 1.8. Here the process output is monitored, and control actions are taken to counteract deviations from required behaviour. In the case of the temperature control system, therefore, the reactant temperature is measured, and if this differs from the desired value the rate of flow of steam is increased or decreased as appropriate to return the temperature towards the desired value. In the case of the motor speed control system, the speed is measured, and the applied voltage modified as required. The effect of feedback is to compensate for *any* discrepancy in the controlled variable, whatever its cause. This type of control strategy is used in such everyday applications as water tank level control using a ball valve, or room temperature control with a thermostat. It is such an important strategy in control engineering that it forms the major subject matter of this book.

Note the distinction between feedforward and feedback, despite the apparent similarities of Figs 1.7 and 1.8. Feedforward involves measuring disturbances directly, whereas feedback measures the controllec variable, and compensates for disturbances only after their *effects* on the controlled variable have taken place. In practice, feedback and feedforward are often combined in a single system.

Summary

Control systems are designed to achieve specified objectives within a given set of constraints. The three common control strategies are open-loop, feedforward and closed-loop control. These strategies are often combined within a single control system. Each strategy requires some model of the process to be controlled.

General characteristics of feedback

2

□ To present a simple description of a closed-loop control system, and hence analyse some of the main properties of feedback.

□ To introduce the concepts of steady-state error, disturbances and disturbance rejection.

□ To show how component characteristics may be linearized about an operating point.

Objectives

In this chapter some of the broad features of feedback control will be analysed using very simple models of control system components. A detailed analysis of closed-loop behaviour must wait until more sophisticated mathematical models have been discussed in Chapter 3. However, it is possible to provide partial answers immediately to such questions as 'how accurate is feedback control?' or 'how well can it counteract disturbances?'

Modelling a feedback loop

Figure 2.1 shows a closed-loop motor speed control system in generalized form, using the symbols almost universally adopted. Here, a voltage corresponding to a measure of the actual speed Ω is compared with a reference voltage, r. The difference between these two voltages is then amplified and applied to the motor, hence generating a control action tending to maintain the speed at a value determined by the reference input.

Let us assume that the motor is modelled by a gain G, the constant of proportionality relating the input voltage to output speed. That is,

$$\Omega = v \times G$$

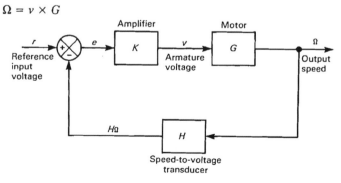

Fig. 2.1 A generalized control loop.

This implies that any change in voltage is reflected immediately in a change in motor speed. No motor can respond instantaneously to a change in applied voltage, of course, so clearly this model is greatly over-simplified. Nevertheless, assuming that the system settles down eventually to some *steady-state* value of motor speed for a given applied voltage, then the above expression can be used to model the *steady-state* condition.

$$G = \frac{\text{steady-state output}}{\text{steady-state input}} = \frac{\Omega_{ss}}{v_{ss}}$$

The entire analysis of this chapter is subject to this important assumption.

Similarly we can model the velocity transducer by another gain. This time, we assume that the transducer produces an output voltage proportional to the speed

$$\text{transducer output voltage} = \text{motor speed} \times H$$

The output voltage from the transducer is compared with a *reference input* voltage r, which is an expression of the desired motor speed. The difference between these two signals – often known as the *error* signal – becomes the input to the amplifier, and the amplifier gain K determines what armature voltage should be applied to the motor. If the speed is too low, the voltage is therefore increased, and vice versa.

The simplest – and very common – form of general closed-loop controller is known as a *proportional controller*, and corresponds to a constant gain, K, acting on the error signal. In the system of Fig. 2.1 the amplifier may be thought of as a proportional controller, producing a control action proportional to the error signal.

Let us assume that the motor control system of Fig. 2.1 has reached a steady state, with the motor running at a constant speed Ω in response to a reference input r volts. In order to assess the performance of the closed-loop system we need to derive a relationship between r and Ω in the absence of any disturbances.

From the figure we can write down immediately

$$e = r - H\Omega$$

and

$$\Omega = KGe$$

Hence

$$\begin{aligned}\Omega &= KG(r - H\Omega) \\ &= KGr - KGH\Omega\end{aligned}$$

Rearranging gives

$$\Omega(1 + KGH) = KGr$$

or

$$\Omega = \frac{KG}{1 + KGH}r$$

Hence the expression $KG/(1 + KGH)$ corresponds to the *closed-loop gain* of the complete feedback system – that is, the factor relating the output (speed) to the reference input in the steady state. The quantity KG is often referred to

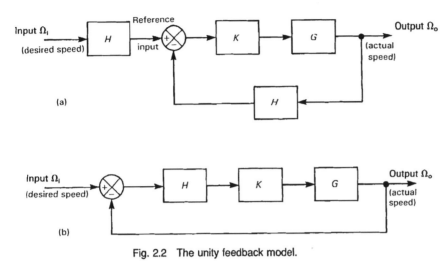

(a)

(b)

Fig. 2.2 The unity feedback model.

as the forward path gain, while KGH is known as the loop gain. Note that the closed-loop gain can therefore be written as

$$\frac{\text{forward path gain}}{1 + \text{loop gain}}$$

It is often more convenient to work with a modified version of Fig. 2.1 in order to obtain an expression directly relating the actual speed to the *desired* motor speed, rather than to a reference input voltage. In this case we can imagine an input 'desired speed' variable, which is then multiplied by a gain exactly equivalent to that of the transducer, in order to give an appropriate reference input voltage, as shown in Fig. 2.2(a). This procedure models the fact that in general the comparison between desired and actual values of a controlled variable will be made in terms of signals representing these measures – an analogue voltage, for example – and not the numerical values themselves. The additional scaling factor H is introduced to reflect this.

Now, it makes no difference whether the gain H is applied before or after the comparator, so long as it is applied to both the signals being compared. Figure 2.2(a) can therefore be re-drawn in the equivalent form of Fig. 2.2(b). This is known as the *unity feedback* form of the closed-loop system, and is an extremely useful concept in the modelling process. Remember, however, that we are assuming here that G and H are pure gains, modelling the steady-state condition. Special procedures are necessary when transducer *dynamics* need to be taken into account, as will be discussed in later chapters.

One further simplification of the unity feedback model can be made. At the design stage it is often convenient to assume that $H = 1$ in Fig. 2.2(b). This allows design calculations – such as determining an appropriate value of controller gain K – to be carried out more simply. When the system is implemented, an appropriately modified value of K can be used, reflecting the various scaling factors involved in the practical system.

The preceding general analysis can now be used to illustrate some of the major features of feedback control. Let us begin by investigating the steady-state error, defined as the difference between desired and actual output when the output has reached a steady, constant value. The unity feedback model of Fig. 2.2(b) makes it particularly easy to relate closed-loop static gain to steady-state error. The error may be calculated easily from the closed-loop gain.

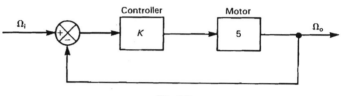

Fig. 2.3

Worked Example 2.1 A motor speed control loop can be modelled in unity-feedback form by Fig. 2.3. Express the steady-state error as a percentage of desired speed if the controller gain K is (i) 1; (ii) 5; and (iii) 10.

Solution The unity-feedback loop with $H = 1$ has a closed loop gain of

$$\frac{\Omega_o}{\Omega_i} = \frac{KG}{1 + KG} \text{ where } G = 5 \text{ in this example.}$$

Hence with the given values of K, the closed-loop system has a gain of $\frac{5}{6} = 0.83$, $\frac{25}{26} = 0.96$ and $\frac{50}{51} = 0.98$, respectively. The percentage error for $K = 1$ is therefore $(1 - 0.83)/1 \times 100\% = 17\%$. Similarly, we have steady-state errors of 4% and 2% approximately for $K = 5$ and 10, respectively.

This example illustrates an important general feature of feedback control systems: assuming that a steady state is eventually reached, the steady-state error is *reduced* as the controller gain (or indeed, the loop gain in general) is *increased*.

Sensitivity of closed-loop gain to changes in parameters

One of the aims of control system design is that the control system should not be too sensitive to changes in the parameters of individual components. After all, as noted in Chapter 1, process parameters can drift with time – or may not be accurately known in the first place. Ideally, a control system should continue to satisfy specifications even if this is the case, so the effects of uncertainties or changes in the parameters of the various components are of great importance. To examine such effects in general terms, consider again Fig. 2.1 and the closed-loop gain relating output speed to input reference voltage.

As derived earlier, the fundamental relationship is

$$\frac{\Omega}{r} = \frac{KG}{1 + KGH}$$

Now, if the loop gain is large, that is, $KGH \gg 1$, this reduces to

$$\frac{\Omega}{r} \simeq \frac{KG}{KGH} = \frac{1}{H}$$

In other words, the precise values of K and G have little effect on the closed-loop gain, *providing the loop gain is sufficiently large*. The feedback loop is relatively insensitive to variations in forward path gain. This is not so for variations in the transducer gain, however. For high loop gain we have

$$\frac{\Omega}{r} \simeq \frac{1}{H}$$

and hence a 10% variation, say in H to a new value of $1.1\,H$, will result in a new closed-loop gain of approximately

$$\frac{\Omega}{r} \simeq \frac{1}{1.1\,H} \simeq \frac{0.9}{H}$$

That is, a change in transducer gain of about 10% leads to a change in closed-loop gain also of about 10%. In general a given (small) percentage variation in feed-back path gain results in a percentage variation in closed-loop gain of approximately the same magnitude (assuming $KGH \gg 1$). This accords with what we might have expected intuitively. The transducer measures the output of the process being controlled, and the control loop cannot reduce error below that introduced by the measuring process itself. So if transducer gain fluctuates by 10%, introducing error into the measurement, the loop cannot compensate, and the effective closed-loop gain also varies by a comparable amount.

Worked Example 2.2

In the system of Worked Example 2.1 suppose that the gain of the motor increases by 10% as a result of reduced friction in the bearings. The transducer gain does not vary. What is the new closed-loop gain for $K = 1$, 5 and 10?

Solution A 10% increase gives a new process gain of 5.5. The closed-loop gains now become:

(a) $K = 1$; $\quad \dfrac{\Omega_o}{\Omega_i} = \dfrac{5.5}{6.5} \simeq 0.85$

(b) $K = 5$; $\quad \dfrac{\Omega_o}{\Omega_i} = \dfrac{27.5}{28.5} \simeq 0.96$

(c) $K = 10$; $\quad \dfrac{\Omega_o}{\Omega_i} = \dfrac{55}{56} \simeq 0.98$

Comparing these with the values derived in Worked Example 2.1 demonstrates the relative insensitivity of the closed-loop gain to changes in forward path gain, and also shows that the closed-loop system becomes progressively *less* sensitive to such forward path parameter variations as the controller gain is increased.

Notice that there is no contradiction between the distinction just made regarding variations in G and H, and the manipulation of the block diagram into unity feedback form in Fig. 2.2. Remember that the additional block of gain H in Fig.

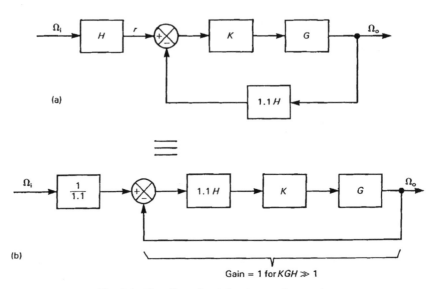

(a)

(b)

Gain = 1 for $KGH \gg 1$

Fig. 2.4   The effect of variation in transducer gain.

2.2(a) reflects the comparison between desired and actual output in terms of another variable – an analogue voltage, say. If transducer gain varies during operation, the scaling factor applied to the input is no longer identical to the actual transducer gain. In the absence of recalibration, the situation becomes as shown in Fig. 2.4(a) for a 10% increase in transducer gain. Attempting to reduce this to unity feedback form leads to Fig. 2.4(b) from which it may readily be seen that, as before, the overall system will suffer a change in closed-loop gain, Ω_o/Ω_i, of about 10%, for high loop gain.

Disturbance rejection

The final general property of feedback control to be considered here is the effect of disturbances. This time let us return to the chemical reactor temperature control system of Fig. 1.2. Major sources of disturbance here might be fluctuations in the temperature or flow rate of steam to the heat exchanger, and changes of

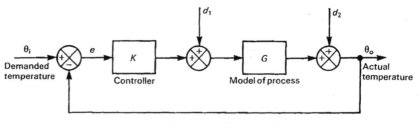

Fig. 2.5   A control loop with disturbance inputs.

12

temperature in the reaction vessel due to the reaction itself, environmental changes, or the addition of new reactants. These can be shown on a block diagram by including new disturbance variables such as d_1, and d_2, as shown in Fig. 2.5. Here the transducer has been modelled as unity gain, the reaction vessel and steam control valve by a combined process gain G, and the controller by a proportional gain K. It is conventional to show *additive* disturbances, as in the figure.

Obtain an expression for the output temperature θ_o in terms of the demanded temperature θ_i, the disturbances d_1, and d_2, and the gains K and G. **Worked Example 2.3**

Solution The output θ_o is made up of three terms:

$$\theta_o = KGe + Gd_1 + d_2$$

where

$$e = \theta_i - \theta_o$$

Hence

$$\theta_o = KG(\theta_i - \theta_o) + Gd_1 + d_2$$
$$\theta_o(1 + KG) = KG\theta_i + Gd_1 + d_2$$

Or

$$\theta_o = \frac{KG}{(1 + KG)}\theta_i + \frac{G}{(1 + KG)}d_1 + \frac{1}{(1 + KG)}d_2$$

The first term is the familiar closed-loop gain expression. The other two terms express the errors introduced into the output as a result of the disturbances. In the absence of the feedback loop the disturbances d_1 and d_2 would give rise to errors of Gd_1 and d_2, respectively, in the output. The factor $1/(1 + KG)$ is therefore a measure of how much the effects of the disturbances are modified by the feedback action. This property is often known as *disturbance rejection*. Note that disturbance rejection is improved by increasing controller gain.

Linearization about an operating point

Although the above analysis is useful, strictly speaking it is only valid if each system element (process, controller, transducer) has an input–output characteristic of the form illustrated in Fig. 2.6(a) – that is, a straight line passing through the origin. It is not uncommon, however, for physical components to possess characteristics which are similar to 2.6(b) or (c). In each case, a single, constant gain cannot be defined for the whole operating range. Often, however, the whole operating range of non-linear components is not needed. The problem can then be overcome by defining new input and output variables in terms of deviations from an operating point, P, as shown in Fig. 2.7 for a characteristic like Fig. 2.6(c). For *small deviations* from the operating point the gain can be defined as the slope of the characteristic at the operating point. Many control systems are designed to maintain the controlled variable at a specific value, and in such cases

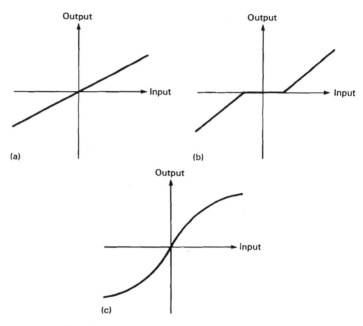

Fig. 2.6   Linear and non-linear static characteristics.

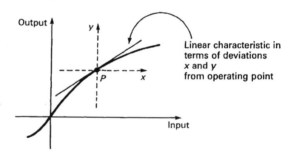

Linear characteristic in terms of deviations *x* and *y* from operating point

Fig. 2.7   Linearization about an operating point.

Non-linearities will not be considered in detail in this book. Many mathematical techniques have been developed to tackle them, however. See, for example: Leigh, J.R., *Essentials of Non-linear Control Theory*, Peter Peregrinus, 1983; Kolk, W.R. and Lerman, R.A., *Non-linear System Dynamics*, Van Nostrand Reinhold, 1992.

linearization as just described is quite valid. If, however, the control system has to operate over a wide range of values of the controlled variable, non-linearities in system components can cause real problems for mathematical analysis.

In control system design and analysis it is normal, in fact, to define *all* variables about an operating point. So, for example, the variables r, e, v and Ω in Fig. 2.1 would all be defined as zero for the motor running at a particular constant operating speed. Values of these variables other than zero represent *deviations* from the operating point. This convention does not affect the mathematical analysis, but makes it easier to apply the mathematical results to actual control systems.

Estimate an appropriate value of gain for the position transducer with the characteristic of Fig. 2.8 at the operating point P shown.

Worked Example 2.4

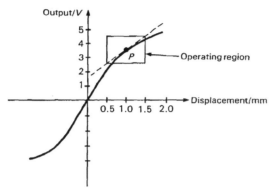

Fig. 2.8

Solution The slope at the operating point is around $2\,\mathrm{V\,mm^{-1}}$ and this is therefore the linearized gain. Note the units of the gain.

Summary

The analysis of this chapter has highlighted a number of important *steady-state* characteristics of feedback.

(1) The steady-state error of a feedback control system (that is, the difference between desired and actual steady-state values of the controlled variable) *decreases* as loop gain is increased.
(2) Fluctuations or uncertainties in *forward path* gain have a relatively small effect on closed-loop gain, providing loop gain is high. The effect of such parameter variations on steady-state closed-loop gain *decreases* as loop gain is increased.
(3) Fluctuations or uncertainties in *feedback path* gain have an effect on closed-loop gain comparable in magnitude to the original changes. Such effects *cannot* be reduced by increasing loop gain.
(4) A feedback system can be relatively insensitive to the presence of disturbances. The effects of such disturbances are *reduced* as controller gain is increased.

Remember that these conclusions follow from a naive model of feedback, in which all system elements are modelled as pure gains. The implications of using more realistic models will be investigated in later chapters.

Finally, it was noted that it is normal to define all variables in a control system in terms of deviations from an operating point. This also enables an appropriate gain to be defined for system elements with characteristics which deviate slightly from strict linearity.

Problems

2.1 The individual elements of the control loop of Fig. 2.9 can be modelled as pure gains. Assuming that the system is stable, calculate the closed-loop gain c/r for (a) $K = 2.5$; and (b) $K = 12$.

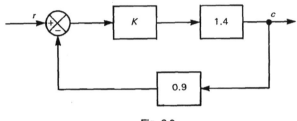

Fig. 2.9

2.2 Obtain an expression for the closed-loop gain of the system of Fig. 2.10.

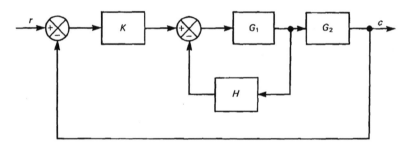

Fig. 2.10

2.3 The output of a certain flowmeter is a differential pressure, p kPa, related to the flow rate q m^{3}s^{-1} by the expression

$$p = 500 q^2$$

over the operating range of the meter.

The meter is to be used to measure flow rates in the range 0.25 to 0.35 m^{3}s^{-1}.

Obtain a linearized expression relating the deviations p' and q' of the differential pressure and flow rate from an operating point corresponding to 0.3 m^{3}s^{-1}.

Modelling dynamic systems   3

Objectives

☐ To outline the control engineering approach to mathematical modelling.
☐ To present a 'library' of standard models of single-input, single-output systems: first-order lag; second-order lag; integrator; and time delay.
☐ To introduce differential equations and frequency response as practical tools for modelling control systems and their components.
☐ To present time and frequency domain models as equivalent descriptions of system behaviour.
☐ To describe the use of normalized step response curves and Bode plots.
☐ To distinguish between system analysis and system identification as routes to a mathematical model.

In Chapter 2 we used the simplest possible models of control system components. Each component was modelled as a steady-state or *static* gain – that is, by the constant of proportionality relating system output and input when both were unchanging with time. Using these simple models it was possible to obtain useful quantitative results about system error, sensitivity, and the ability to counteract steady disturbances. To do this, however, we chose to neglect the way the output (or, indeed any other system variable) changed with time in response to a time-varying input. In other words, the models included no information whatsoever about system dynamics. Furthermore, we also tacitly assumed that the overall system would be *stable* – that is, the system output would not increase indefinitely with time so long as neither the input nor any disturbance did so.

Now, one of the major tasks of the control engineer is to design systems whose *dynamic* behaviour falls within the limits given in the system specification. Not only must the overall system always be stable, but the way it behaves during the *transient* period of 'settling down' after a change in input or disturbance must also be carefully controlled. In order to do this, an adequate mathematical model of system dynamics is required.

The modelling approach

No mathematical model is ever 'correct', in the sense of offering a perfect input–output description of system dynamic behaviour. Fortunately, however, control engineers are not primarily interested in such 'perfect' input–output descriptions. The most important function of a model for control purposes is as a basis for designing a control system which will conform to given specifications in spite of the inherent limitations of the model. It is in these terms that the adequacy or otherwise of a mathematical model must be judged. The fine details of the model, in fact, will often depend as much on the control system specification as on the physical nature of the system being modelled: for a given system, the degree of sophistication required in a mathematical model is influenced by the performance being demanded from the final overall design.

Control engineers tend to rely on a comparatively small number of standard models which have proved to be particularly useful idealizations of features commonly encountered in dynamic systems. Such standard models can be thought of as offering the designer a 'library' of mathematical tools. The art of mathematical modelling is very often that of choosing and combining appropriate standard models in the light of the system specification and the designer's experience. Sometimes the modelling will call for the analysis of the behaviour of individual system components, using a knowledge of the laws of physics, for example. In other cases it may be more appropriate to carry out experimental tests before selecting a standard model with appropriate characteristics. In practice, both analysis and experiment are often necessary.

One important advantage of using standard models is that they make it easier to carry over experience with one system to others with similar dynamic characteristics. Control engineers are called upon to model a wide range of physical systems – from entire chemical plant to computer disk drives. The fact that a common approach can be adopted to control system design in such apparently widely different circumstances owes much to the use of standard models of the type described in this chapter.

A first-order differential equation model

One type of model commonly used by control engineers takes the form of a differential equation relating system input and output. Including derivatives with respect to time in the model brings in the required dynamic information -- that is, information about the way the system variables change with time. As a first example, let us return to the motor speed control system of Chapter 1, but this time we shall assume that both input voltage v and output speed Ω can vary generally with time, leading to the block diagram of Fig. 3.1. In order to derive a differential equation model relating input and output, we need to look in more detail at how such motors function. First, though, one further comment is called for.

In the course of the following derivation, various modelling assumptions will be made. This is necessary in order to obtain a final mathematical model of a type which can be handled without too much difficulty. But there is also another important point about such modelling assumptions, which is not always appreciated. Since assumptions *have* to be made to avoid intractable mathematics, devices such as transducers and actuators are carefully engineered so that the assumptions hold as well as possible in practice. We could perhaps speak of a gradual convergence of mathematical models and hardware performance as

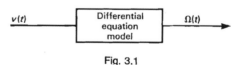

Fig. 3.1

18

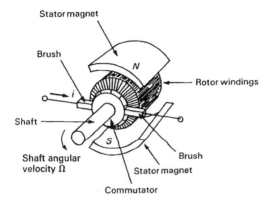

Fig. 3.2 An armature controlled d.c. motor.

designs have been refined over the years. So, in the derivations of the models of this chapter, bear in mind that modelling assumptions are not made in isolation, but in the light of experience, and with the ultimate end of the modelling exercise always in view. Now let us turn to the motor.

Figure 3.2 shows, in idealized form, part of a small d.c. servo-motor of the type used in speed and position control systems. The applied voltage $v(t)$ causes a current $i(t)$ to flow in the armature winding. Because the armature is in a constant magnetic field, a torque is developed which is proportional to the armature current. Hence we can write

$$\text{torque} = K_m i(t)$$

where K_m is a constant of proportionality known as the motor torque constant.

In the absence of friction all the generated torque is available to accelerate the motor shaft and the attached load according to Newton's Second Law of Motion:

$$\text{available torque} = \text{moment of inertia} \times \text{angular acceleration}$$

Assuming negligible friction, therefore (another important modelling assumption), we have:

$$K_m i(t) = J \times \frac{d\Omega(t)}{dt}$$

where J is the moment of inertia of the motor and load.

This expression is itself a differential equation modelling the dynamics of the motor. However, we require the motor velocity Ω to be expressed in terms of the system input v, rather than the current i flowing in response to the applied armature voltage.

Since the armature is a conductor rotating in a magnetic field, a voltage will be induced in the winding so as to *oppose* the flow of current. This *back e.m.f.*, e, is proportional to the speed of rotation giving

$$e = K_b \Omega$$

where K_b is a second constant of proportionality.

When modelling a system we cannot change, we aim for a model of a type we can handle; when building a device to be used for control purposes, we often try to design it to behave like a suitable ideal model.

This is a typical example of the convergence of modelling decisions and hardware design. Armature-controlled d.c. servo motors are carefully designed so that this expression closely models the behaviour of the motor.

From now on I shall represent $i(t)$ simply by i, $d\Omega/dt$ by $\dot{\Omega}$, and so on.

Remember that both v and e may vary generally with time.

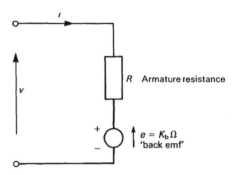

Fig. 3.3 Motor equivalent circuit.

In many applications, d.c. servo-motors can be modelled by the equivalent circuit of Fig. 3.3. (The major modelling assumption is that the effect of armature inductance can be neglected in comparison with that of its resistance.) The effective voltage across the armature resistance is therefore $v - e$, and Ohm's law for the circuit can be written

$$v - e = iR$$

Combining this with the torque equation gives

$$\frac{K_m}{R}(v - K_b\Omega) = J\dot{\Omega}$$

Rearranging, we have

In general, a differential equation may involve the derivatives with respect to time of either input or output or both. It is quite common in control engineering, however, to meet system models which involve derivatives of the *output* variable only, as in this case.

$$\frac{K_m}{R}v = J\dot{\Omega} + \frac{K_mK_b}{R}\Omega$$

or

$$\frac{JR}{K_mK_b}\dot{\Omega} + \Omega = \frac{1}{K_b}v$$

This is a first-order linear differential equation model relating motor input v and output Ω.

Worked Example 3.1   (a) Using the notation of the previous equation, what is the steady-state gain of the motor?
(b) What are the dimensions of the coefficient of $\dot{\Omega}$?

Solution
(a) In Chapter 2 the steady-state gain of the motor was defined as

$$\frac{\text{steady-state output}}{\text{steady-state input}} = \frac{\Omega_{ss}}{v_{ss}}$$

Under steady-state conditions $\Omega = d\Omega/dt = 0$, since the velocity does not vary with time. Hence

$$\Omega_{ss} = \frac{1}{K_b} v_{ss}$$

and the steady-state gain is $1/K_b$
(b) The coefficient of $\dot{\Omega}$ is JR/K_mK_b. The dimensions of Ω are $[\text{TIME}]^{-1}$ so the dimensions of $\dot{\Omega}$ are $[\text{TIME}]^{-2}$. Each term on the left-hand side of the equation must have dimensions of $[\text{TIME}]^{-1}$, so the dimensions of the co-efficient of $\dot{\Omega}$, JR/K_mK_b, must be $[\text{TIME}]$.

The results of the previous example allow us to make a major simplification of the way we write the system differential equation. Since each term in the coefficient of $\dot{\Omega}$ is a constant, the coefficient itself is a constant. And because it has the dimensions of $[\text{TIME}]$ it is known as the *time constant* of the system and given the symbol τ. We can therefore re-write the differential equation in the engineering *standard form*:

$$\tau\dot{\Omega} + \Omega = kv$$

where k is the steady-state gain.

Writing a particular first-order equation of this type in standard form enables the values of steady-state gain and time constant to be obtained simply by inspection.

A motor can be modelled by the differential equation

$$\dot{\Omega} + 0.5\Omega = 2v$$

where Ω is expressed in $\text{rad}\,\text{s}^{-1}$ and v in volts.

(a) What is its time constant?
(b) An input voltage of $5\,\text{V}$ is applied to the motor. What is the steady-state output velocity predicted by the differential equation model?

Worked Example 3.2

Solution
(a) The differential equation must first be written in standard form – that is, the coefficient of Ω must be unity. Dividing each side of the equation by 0.5 gives

$$2\dot{\Omega} + \Omega = 4v$$

Comparing this with the standard form

$$\tau\dot{\Omega} + \Omega = kv$$

we see immediately that $\tau = 2\,\text{s}$.
(b) In the steady state $\dot{\Omega} = 0$ and $\Omega_{ss} = kv_{ss}$.
From the standard form of the equation $k = 4$. In this example $v_{ss} = 5\,\text{V}$. Hence

$$\Omega_{ss} = 4 \times 5$$
$$= 20\,\text{rad}\,\text{s}^{-1}$$

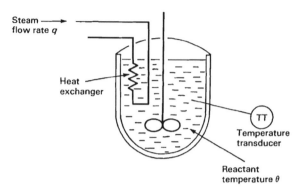

Fig. 3.4   A chemical reactor.

Writing a system differential equation in terms of the *standard parameters* τ and k results in a much more generally applicable modelling tool. It is the first model in our 'library', in fact. Equations of this form have proved to be extremely useful in modelling a wide range of systems whose physical character varies considerably. For example, consider the chemical reactor first introduced in Chapter 1, and illustrated in more detail now in Fig. 3.4. For many systems of this type an approximate model of the relationship between input (steam flow rate q) and output (reactant temperature θ) takes the form

$$\tau\dot{\theta} + \theta = kq$$

that is, a first-order differential equation again characterized by a time constant τ and a steady-state gain k, very similar to the motor example.

Characterizing first-order dynamics: the step response

Although I have stressed the importance of dynamics to the control engineer, I have not yet considered the specific dynamic characteristics implied by our differential equation model of motor behaviour

$$\tau\dot{\Omega} + \Omega = kv$$

In theory, we could use standard techniques for solving this equation to determine the way Ω varies with time for any given input v. However, the aim of control system design is to ensure that the system will behave properly *regardless* of the way the input varies with time – and all possible time variations of the input cannot be known in advance. It is therefore usual to take the form of system response to a standard 'test' input as characteristic of the particular system. One of the most common forms of standard input used in control is the *step input*, in which the input to the system is changed as quickly as possible from one constant value to another. The way the system output responds to this test input is known as *the step response*. Step response testing is relatively easy to carry out in practice, and completely characterizes the system. Since the input is changed to a new value, the steady-state value of the output will indicate the static gain of the

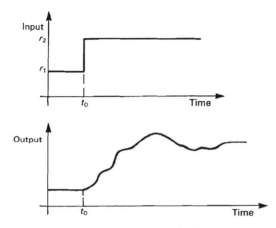

Fig. 3.5 Step-response testing.

system. In addition, the way the output changes before reaching its steady-state value gives information about the *transient behaviour* of the system.

The general idea is illustrated in Fig. 3.5 for a step change in input from r_1 to r_2 at time t_0. To simplify matters, step response is usually referred to step change in input variable from zero to one, occurring at time $t = 0$.

Adopting this convention, then, the theoretical step response of the motor can be found by solving the differential equation for the unit step input

$$v = 0, t < 0$$
$$v = 1, t \geq 0$$

We assume that $\Omega = 0$ before the step input is applied and solve

$$\tau\dot{\Omega} + \Omega = k, t \geq 0$$

The solution to this equation is well known (it can be obtained by separating variables and integrating). After substituting initial and final conditions it can be written in terms of the steady-state output Ω_{ss}:

$$\Omega = \Omega_{ss}\left[1 - \exp\left(\frac{-t}{\tau}\right)\right]$$

This analytic expression describing the step response is plotted in *normalized* form in Fig. 3.6. The system output is given in terms of the steady-state value, and the time axis is labelled in terms of the time constant of the system. Important features to note are:

(i) The output reaches about 63% of its final value after a time equal to 1 time constant.
(ii) The initial slope of the response is $1/\tau$.

Remember that input and output variables are defined as deviations from a steady-state operating point, as described in Chapter 2. So in the case of the motor a value of $v = 0$ does not imply zero armature voltage and hence zero speed, but that the actual input voltage and output speed correspond to the operating point.

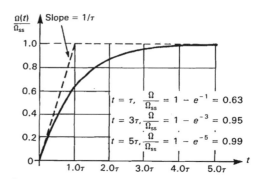

Fig. 3.6 Normalized first-order step response curve.

Worked Example 3.3 Using the differential equation model of Example 3.2 and the curve of Fig. 3.6,

(a) What is the speed of the motor 2 s after applying an input step of 1 V?
(b) How long does it take for the motor to reach 95% of its steady-state value?

Solution

(a) From Example 3.2 we know that $\tau = 2\,\text{s}$. From Fig. 3.6 it can be seen that after one time constant the output reaches $0.63\Omega_{ss}$ in response to a unit input step. In this case, $\Omega_{ss} = kv_{ss} = 4$, so after two seconds $\Omega = 0.63 \times 4 = 2.52\,\text{rad s}^{-1}$.

(b) Figure 3.6 indicates that 95% of the steady-state value will be reached after about 3 time constants. In this case, then, after about 6 s.

This 'scaling' property is a characteristic of models based on linear differential equations. For a full discussion of the meaning of the term 'linear' in this sense, see Meade, M.L. and Dillon, C.R., *Signals and Systems*, Van Nostrand Reinhold, 1985 (2nd edition 1991).

In the previous example the input step was 1 V. However, the normalized curve of Fig. 3.6 may be used for step inputs of other magnitudes, simply by scaling the output accordingly (providing the differential equation model continues to hold). For example, if the input step had been 2 V, then the output velocity after 2 s would have been $2 \times 2.52 \simeq 5\,\text{rad s}^{-1}$.

The easiest way to calculate specific step responses is to use a normalized step response curve. However, it is worth memorizing the main features of such standard models, and you should aim at being able to reproduce a sketch similar to Fig. 3.6. Incidentally, this curve illustrates how the output of a system of this type 'lags behind' a change in input. For this reason this standard model is known as a *first-order lag*.

We have derived the first-order lag model by analysing in detail the dynamic behaviour of a single component – a d.c. motor. However, standard models like this can also be used to describe the input–output behaviour of complete systems, where the systems can consist of many individual components. In fact, we do not even need to test or analyse individual components in order to obtain a mathematical model of a complete system. The following example illustrates this in the context of a simple closed-loop control system.

Worked Example 3.4 The measured response of the position control system of Fig. 3.7 resulting from a 10 mm step change in demanded position is shown in Fig. 3.8. Estimate the time

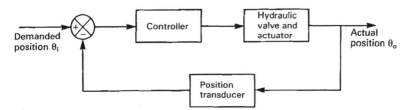

Fig. 3.7 A position control system.

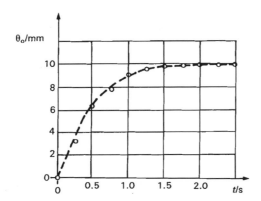

Fig. 3.8 Response to a step change in demanded position

constant and steady-state gain of the closed-loop system, and write down a differential equation modelling the system dynamics.

Solution An estimate of the time constant can be made by determining how long the system output takes to reach 63% of its steady-state value. The steady-state output is 10 mm; from Fig. 3.8, a value of 6.3 mm is reached after approximately 0.5 seconds. (The curve deviates somewhat from an 'ideal' first-order response, but $\tau = 0.5$ is a reasonable estimate.) The steady-state gain is $10/10 = 1$, so the differential equation is

Note that a steady-state gain of unity in this case corresponds to zero steady-state error.

$$0.5\dot{\theta}_o + \theta_o = \theta_i$$

An alternative description of system behaviour: frequency response

Suppose that a sinusoidally-varying voltage is supplied to the d.c. motor of the previous section. How will the motor velocity vary with time? This might appear to be a rather contrived question, yet behind it lies one of the most important modelling and design tools of control engineering.

Intuitively it seems likely that the motor speed will itself vary periodically in some way in response to an applied sinusoid. To obtain a precise answer, however, we need to interpret the system differential equation in the light of two important properties of sinusoids. The first property of interest is that a sinusoid can be differentiated indefinitely, each differentiation resulting in another sinusoid. For example

$$\sin \omega t \xrightarrow{\frac{d}{dt}} \omega \cos \omega t \xrightarrow{\frac{d}{dt}} -\omega^2 \sin \omega t \to \cdots$$

Remember that $\cos \omega t$ is also sinusoidal and can be written $\sin(\omega t + \frac{\pi}{2})$.

The second important characteristic is that any two sinusoids at the same frequency may be added together to give a third. This resultant sinusoid generally differs in amplitude and phase from the first two, but is of the same frequency.

Armed with a knowledge of these two properties of sinusoids, we can make the following observation.

If the motor voltage v varies sinusoidally, according to the expression $v(t) = V \sin \omega t$, then in order to satisfy the differential equation

$$\tau \dot{\Omega} + \Omega = kv$$

both Ω and $\dot{\Omega}$ will also be sinusoidal in nature. Then the left-hand side can be combined into a single sinusoid, which must correspond to the applied sinusoidal voltage multiplied by the gain k.

By a 'perfect' sinusoid, I mean one which has existed and will continue to exist for all time. This is what is implied when we write down an expression of the form $v(t) = V \sin \omega t$. We can approximate this in practice by taking measurements only when sufficient time has elapsed after switching on the input. A good rule of thumb for first-order systems is to allow a time equivalent to 5 time constants before taking measurements of phase shift and amplitude ratio.

This is a fundamental result, which applies to any linear differential equation with constant coefficients. It means that the theoretical response of the system to a perfect sinusoidal input is a perfect sinusoidal output. The input and output generally differ in amplitude and phase but are at the same frequency, as illustrated in Fig. 3.9. This 'frequency preservation' property is characteristic of linear differential equation models, and is often obeyed sufficiently well in practice to be a useful test method. In such *frequency response testing* a sinewave is applied to the input of the system under test and the output is allowed to settle down to a steady-state sinusoidal response. The amplitude and phase shift of the output with respect to the input are then measured. This is repeated for various frequencies over the range of interest. Special instruments such as *frequency response analysers* are available for this purpose.

In a real physical system, of course, the steady-state system output will not be a pure sinusoid and it is also likely to be contaminated by noise, as illustrated in

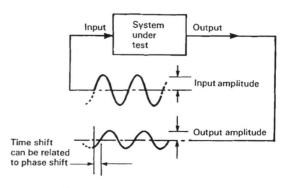

Fig. 3.9 Frequency response testing.

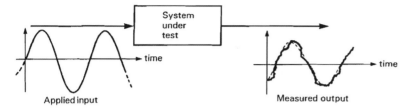

Fig. 3.10

Fig. 3.10. Providing the approximation is sufficiently close, however, the model is still useful. All mathematical models are only ever approximations to actual system behaviour, and their limits of validity must always be borne in mind.

The frequency response function

The change in amplitude and phase introduced by the system can be related analytically to the corresponding differential equation, most conveniently by using *phasors*.

A more thorough discussion of phasors is given in Szymanski, J.E., *Basic Mathematics for Electronic Engineers*, Chapman & Hall, 1989.

The properties of phasors are summarized as follows:

(1) The sinusoid $v(t) = V\sin(\omega t + \phi)$ is represented on the *phasor diagram* of Fig. 3.11 by a line of length V making an angle ϕ with the phase reference axis. V is the amplitude of the sinusoid, ϕ the phase, and ω is its angular frequency in $\mathrm{rad\,s^{-1}}$.

(2) Phasors can be manipulated algebraically using complex number notation. Hence the phasor corresponding to $v(t)$ is, in polar form, $V = |V|e^{j\phi}$. $|V|$ is the phasor magnitude, and ϕ its phase angle. Note that the phasor gives no information about the frequency of the sinusoid.

(3) Multiplication of the phasor $V = |V|e^{j\phi}$ by the complex *phasor operator* $G = |G|e^{j\theta}$ results in another phasor, with magnitude $|G| \times |V|$ and phase angle $\theta + \phi$.

(4) As a special case, rotation of a given phasor anticlockwise by $\frac{\pi}{2}$ rad or 90° corresponds to multiplication by the phasor operator j. Multiplication by j is equivalent to increasing the phase by $\frac{\pi}{2}$ rad (90°) leaving the magnitude unchanged.

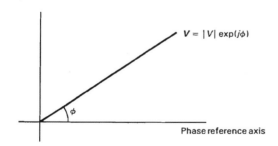

Fig. 3.11

27

(5) Differentiation of the sinusoid $f(t) = \sin \omega t$ gives $df/dt = \omega \cos \omega t = \omega \sin(\omega t + \frac{\pi}{2})$. Differentiation therefore corresponds to multiplication of the associated phasor by the phasor operator with magnitude ω and phase $\frac{\pi}{2}$ rad (90°) − that is, by the operator $j\omega$.

Using the standard rules of phasor manipulation we can immediately transform the differential equation

$$\tau\dot{\Omega} + \Omega = kv$$

into a *phasor equation*. Writing the input phasor as V and the output phasor as Ω we have

$j\omega\Omega$ is the phasor corresponding to $\dot{\Omega}$.

$$j\omega\tau\Omega + \Omega = kV$$

Hence

$$\Omega = \left[\frac{k}{1 + j\omega\tau}\right]V$$

The term $k/(1 + j\omega\tau)$ is a complex phasor operator, having a unique magnitude and phase angle for each value of frequency, ω. At any given frequency its magnitude, $|k/(1 + j\omega\tau)|$, represents the amplitude ratio of the system − that is, the factor by which the amplitude of the output sinusoid differs from that of the input. Its angle represents the phase shift introduced by the system. The full complex expression is usually known as the *frequency response function* or *frequency transfer function* and is represented symbolically as a function of $j\omega$ − thus

$$\frac{\Omega}{V} = G(j\omega) = \frac{k}{(1 + j\omega\tau)}$$

Frequency-domain models are important in control engineering not because the frequency responses of components or systems are of particular interest in themselves, but because they lead to useful models for control system design. The frequency-domain representations developed in this chapter will come into their own when we go on to look at design in later chapters.

* It is unusual in control engineering texts to distinguish between upper- and lower-case Greek symbols. In such cases it should be clear from the context whether a time or frequency-domain variable is meant.

We now have two, absolutely equivalent models of the motor. The differential equation model, often known as a *time-domain* model, describes how a general time-varying input voltage is 'processed' by the motor to give a time-varying output velocity. The frequency response or *frequency-domain* equivalent, on the other hand, describes how a steady-state input sinusoid is 'processed' to produce a steady-state output sinusoid.

The two equivalent first-order lag models are illustrated in block diagram form in Fig. 3.12. Note the convention of using lower-case symbols in the time domain, and upper case in the frequency domain*.

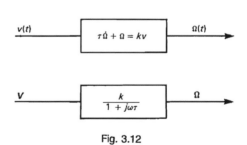

Fig. 3.12

Bode plots

Just as a normalized step response curve was the most convenient way of characterizing the time-domain behaviour of a first-order lag, so normalized frequency response curves play a similar role in the frequency domain. The shape of these curves can be obtained from the frequency response function itself. First, let us consider the limiting cases of the magnitude $|G(j\omega)|$, where

$$|G(j\omega)| = \left| \frac{k}{1 + j\omega\tau} \right| = \frac{k}{\sqrt{1 + \omega^2\tau^2}}$$

When ω is very small compared with $1/\tau$ – that is, when $\omega\tau \ll 1$ – then $|G(j\omega)| \simeq k$, which is the steady-state or *low frequency gain*.

When $\omega\tau \gg 1$, on the other hand, $|G(j\omega)| \simeq k/\omega\tau$ – that is, the magnitude is inversely proportional to frequency at sufficiently high frequencies.

The phase shift for these limiting cases can also be evaluated from the frequency response function. Using the normal rules of complex number arithmetic,

$$\angle G(j\omega) = \text{(phase angle of numerator)} - \text{(phase angle of denominator)}$$
$$= 0 - \arctan(\omega\tau)$$

This is usually written as $\phi(\omega) = -\arctan(\omega\tau)$.
For $\omega\tau \ll 1$, therefore, the phase shift $\phi \simeq -\arctan 0 = 0°$, while
For $\omega\tau \gg 1$, $\phi = -\arctan(\omega\tau) \simeq -90°$.

> This agrees with our earlier definition of steady-state gain, since the limit as $\omega \to 0$ may be thought of as representing a constant input.

Evaluate $|G(j\omega)|$ and $\phi(\omega)$ for the frequency $\omega_c = 1/\tau$ **Worked Example 3.5**

Solution For $\omega_c = 1/\tau$ we have $\omega_c\tau = 1$ and $G(j\omega) = k/(1 + j1)$

$$|G| = \left| \frac{k}{1 + j1} \right| = \frac{k}{|1 + j1|} = \frac{k}{\sqrt{1 + 1}} = \frac{k}{\sqrt{2}}$$

$$\phi = -\arctan(\omega_c\tau) = -\arctan(1) = -45°$$

It is usual to plot $G(j\omega)$ as separate graphs of amplitude ratio $|G(j\omega)|$ and phase shift $\phi(\omega)$. Plots are most conveniently used in their *normalized form*: amplitude ratio is normalized with respect to low frequency gain k, and frequency is normalized with respect to the cut-off frequency $\omega_c = 1/\tau$. This means that $|G/k|$ is plotted against $\omega\tau$ (or ω/ω_c), as illustrated in Fig. 3.13. The amplitude ratio plot therefore indicates the factor by which $|G|$ at any particular frequency differs from k, while the use of a frequency axis labelled as multiples of the cut-off frequency ($\omega_c = 1/\tau$) enables systems with widely differing time constants to be described in a uniform way. Logarithmic amplitude ratio and frequency axes are usually employed as shown in Fig. 3.13. The vertical axis of the $|G|/k$ plot has been labelled as both a ratio and in decibels, where the decibel measure is defined as $20\log_{10}(|G|/k)$.

Frequency response plots in this form are known as *Bode plots*.

For many purposes the asymptotic, straight-line, approximations shown as dotted lines are sufficiently accurate. The important features of the Bode plot of a first-order lag are as follows:

> A knowledge of the decibel measure is assumed in this text. For details see: O'Reilly, J.J., *Telecommunications Principles*, Van Nostrand Reinhold, 1984.

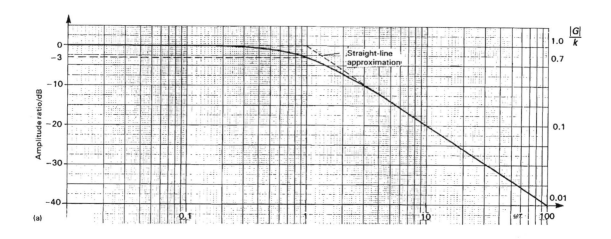

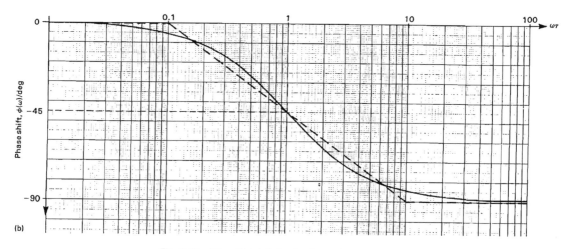

Fig. 3.13 Normalized first-order frequency response curves.

Bode was one of the pioneers of feedback system design in the USA in the 1930s and 1940s, and the first to use frequency response plots in this form. Bode developed his ideas in the context of feedback amplifiers; they were subsequently applied with great success to control systems, forming the basis of the approach covered in Chapter 4.

(1) The amplitude ratio at $\omega_c = 1/\tau$ is $1/\sqrt{2}$ or $-3\,\mathrm{dB}$ of its low frequency value. This frequency is variously known as the cut-off frequency, the 3 dB point, the breakpoint, or the corner frequency. The phase shift at this frequency is $-45°$.

(2) As seen earlier, for $\omega\tau \gg 1$, $|G(j\omega)| = k/\omega\tau$. The amplitude ratio is inversely proportional to frequency, and therefore at sufficiently high frequencies it decreases (*rolls-off*) by a factor of 10 for each 10-fold increase in frequency. On logarithmic axes this results in the straight-line asymptote of Fig. 3.13. A factor of 10 is equivalent to $20\log(10) = 20\,\mathrm{dB}$, so the roll-off is usually referred to as 20 dB per decade.

(3) At low frequencies ($\omega \ll 1/\tau$) the phase shift is $0°$; at sufficiently high frequencies ($\omega \gg 1/\tau$) it approaches $-90°$.

A temperature-to-voltage transducer can be modelled by the differential equation

Worked Example 3.6

$$0.8\dot{v} + v = \theta$$

where v is the output voltage and θ the input temperature. What is the amplitude ratio (in dB) and phase shift of the transducer at a frequency of 1 Hz?

Solution   The differential equation model corresponds to a time constant of 0.8 s and a low frequency gain of 1. To use the standard curves, the frequency of 1 Hz must first be expressed in rad s^{-1}, and then converted to a normalized frequency in terms of $\omega\tau$.

1 Hz corresponds to $\omega = 2\pi$ rad s^{-1}. For $\tau = 0.8$ the normalized frequency $\omega\tau$ is therefore $2\pi \times 0.8 \simeq 5$. (That is, 1 Hz is about 5 times the cut-off frequency.) At this frequency the amplitude ratio is -14 dB, while the phase shift is about $-80°$.

This section and the previous one have dealt in some detail with the first-order lag model, which is a useful idealization of a number of commonly-occurring physical systems. Many systems, however, have properties which cannot be characterized adequately using such a model. In the remainder of this chapter a number of other simple standard models will be introduced, applying the time and frequency-domain techniques just discussed. All the models are applicable to many systems other than the examples presented.

An integrator model

Figure 3.14 shows a hydraulic cylinder of the type often used in position control systems. If we define system input as fluid flow rate, and output as piston displacement, then it is clear that the output will continue to increase even while the input is held constant. Furthermore, the output will remain at a constant value, usually other than zero, when the input is zero. The hydraulic cylinder can be thought of as representative of a class of systems with these general input–output characteristics, and by idealizing this feature, we can derive another standard model for our library. Returning to our specific example, if we neglect friction, fluid compressibility, leakage past the piston and inertia, the velocity of the piston is directly proportional to the flow rate of hydraulic fluid into the cylinder. (Fluid flow rate, and hence piston displacement, can be controlled by means of a suitable valve.) The differential equation relating flow rate q and piston displacement x is therefore:

$$\frac{dx}{dt} = Kq$$

where K depends on the cross-sectional area of the cylinder and can be thought of as a gain. Assuming that $x = 0$ at time $t = 0$, we can write:

$$x = K\int_0^t q \, dt$$

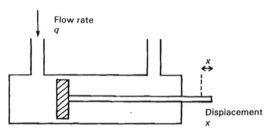

Fig. 3.14 An idealized hydraulic cylinder.

This model is known as an integrator, since the output of the system is directly proportional to the time integral of the input.

Worked Example 3.7

(a) Sketch the step response of the integrator model of the actuator – that is, the time variation of x when q is changed instantaneously from 0 to 1 unit.
(b) Derive the frequency response function of the integrator.

Solution (a) The step response is shown in Fig. 3.15 assuming $x = 0$ initially. Note that the output position increases indefinitely with time, the slope of the line being K. We can show this from the integral version of the system differential equation. If

Of course, the displacement cannot increase indefinitely with time – only within the physical limits of the cylinder. The integrator model therefore holds only within these limits. Within similar limits, integrator models are useful for many other systems whose output tends to increase at a constant rate in response to a steady input.

$$\frac{dx}{dt} = Kq$$

then

$$x = K \int q \, dt$$

If $q = 1$ for $t > 0$, then

$$x = K \int dt = Kt + c$$

where c is an arbitrary constant equal to zero if $x = 0$ at $t = 0$.

(b) The frequency response function $G(j\omega)$ can be derived using the phasor approach as before. The differential equation model of the system is

$$\frac{dx}{dt} = Kq$$

Using phasor notation and the $j\omega$ differentiation operator we have

$$j\omega X = KQ$$

$$X = \frac{K}{j\omega}Q$$

32

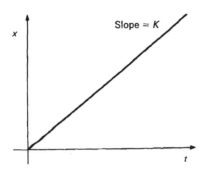

Fig. 3.15 Integrator step response.

Hence

$$G(j\omega) = \frac{X}{Q} = \frac{K}{j\omega}$$

Divide output phasor by input phasor to obtain the frequency response function.

Again, an integrator model may be characterized by its normalized Bode plot of $|G|/K$ and ϕ against frequency. (This time frequency does not have to be normalized with respect to a time constant.) This Bode plot is shown in Fig. 3.16. The particular features to note in the integrator plots are:

(1) A constant phase shift of $-90°$ and an amplitude ratio slope of $20\,\text{dB}$ per decade for all frequencies. The amplitude ratio is $0\,\text{dB}$ for $\omega = 1\,\text{rad}\,\text{s}^{-1}$.
(2) An amplitude ratio which increases indefinitely as the frequency decreases.

Worked Example 3.8

A particular hydraulic system can be represented schematically by Fig. 3.17, where the displacement of the vane actuator is controlled by the electrical signal to the servovalve. The valve can be represented by a pure gain of $6.36 \times 10^{-3}\,\text{m}^3\,\text{s}^{-1}\,\text{A}^{-1}$, relating fluid flow rate to servovalve current. The actuator can be modelled as an integrator. Manufacturer's literature indicates that $1.56 \times 10^{-4}\,\text{m}^3$ of fluid needs to flow through the actuator to turn it through an angle of one radian.

(a) Derive the frequency response function relating valve current i to actuator displacement θ.
(b) Sketch the corresponding Bode plot.

Solution From the data given we can write:

$$q = 6.36 \times 10^{-3} i$$

and

$$q = 1.56 \times 10^{-4} \dot{\theta}$$

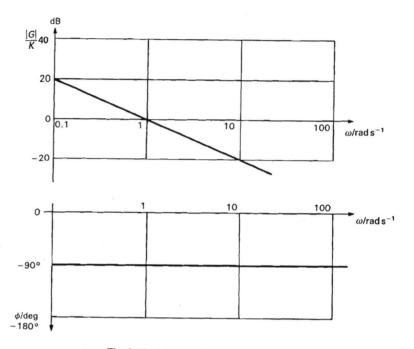

Fig. 3.16 Integrator frequency response.

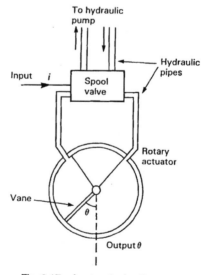

Fig. 3.17 A rotary hydraulic actuator.

Hence

$$\dot{\theta} = \frac{6.36 \times 10^{-3}}{1.56 \times 10^{-4}} i = 41i$$

To convert this differential equation into a phasor equation, replace i by I and $\dot{\theta}$ by $j\omega\theta$. Hence

$$j\omega\theta = 41I$$

and

$$G(j\omega) = \frac{\theta}{I} = \frac{41}{j\omega}$$

This is an integrator model, whose corresponding Bode plot is shown in Fig. 3.18. Note that the amplitude ratio $|G| = 1$ ($0\,$dB) when $\omega = 41$, and that $|G| = 41$ when $\omega = 1$. Converting the latter value to decibels gives

$$\begin{aligned} |G| &= 20\log_{10}41\,\text{dB} \\ &= 32\,\text{dB} \end{aligned}$$

at $\omega = 1$. The Bode plot of Fig. 3.18 is therefore identical to that of Fig. 3.17, but raised by $32\,$dB corresponding to the gain of 41. As in Fig. 3.17, the phase shift is $-90°$ for all frequencies.

Incidentally, it is worth remembering that the amplitude ratio corresponding to an integrator model

$$G(j\omega) = \frac{K}{j\omega}$$

is $0\,$dB at a value of ω numerically equal to K. Integrator models are also often used in the form

$$G(j\omega) = \frac{1}{j\omega T_i}$$

where T_i is a constant. In this case $|G|$ becomes $0\,$dB when $\omega = 1/T_i$. Integrator models will be used in this form in Chapter 4.

A second-order lag model

Many physical systems exhibit oscillatory behaviour in response to a sudden disturbance from equilibrium: car suspension systems, for example, or some mechanical structures in response to wind-loading. Clearly neither the first-order lag nor the integrator model reflects this sort of behaviour. To represent such oscillatory responses by a differential equation, in fact, we need to include second- (or higher-) order derivatives. One example must serve here as representative of a whole range of such second-order systems: the dynamics of a radio telescope antenna structure. The telescope drive system can be represented schematically by Fig. 3.19. We have already looked at a typical representation of electric motor dynamics, but how will the antenna structure itself respond to disturbances from equilibrium? In particular, how will the angular displacement

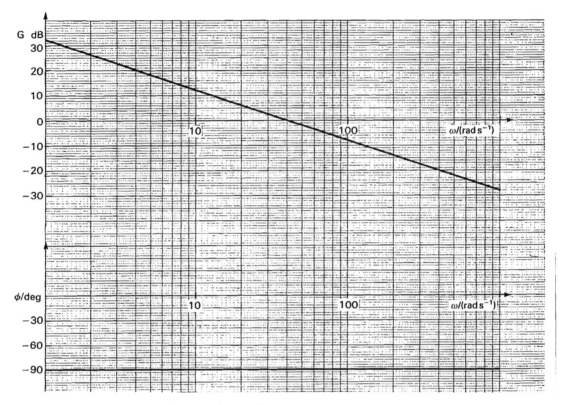

Fig. 3.18 Actuator frequency response.

of the structure be related to the angle of the drive shaft from the gearing? Clearly, under dynamic conditions the two angles θ_i and θ_o will not always be identical, since the structure can twist under load.

The most common way of modelling the dynamics of a complex structure such as this is by means of a so-called lumped parameter model, in which the distributed effects of friction, 'springiness' (stiffness) and so on throughout the structure are modelled as discrete components.

Figure 3.20 illustrates these lumped friction and stiffness 'components' for the telescope. Friction itself is a complicated physical phenomenon. However, its main characteristic is clearly that it opposes the motion of the telescope, and that it increases as the speed of rotation of the antenna increases. The simplest linear model is therefore that the frictional torque is proportional to the angular velocity of the antenna $\dot{\theta}_o$ and acts in the opposite direction to this velocity. Hence we write

friction torque $= -k_f\dot{\theta}_o$, where k_f = constant

36

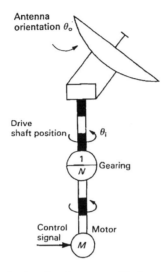

Fig. 3.19 A radio telescope positioning system.

The fact that the structure is not infinitely stiff, but can twist in response to applied torques, is modelled by the *lumped torsional stiffness*. In this model

twisting torque $= k_s(\theta_i - \theta_o)$, where $k_s =$ constant

This is the angular equivalent of Hooke's law.

The torque available to accelerate the antenna will be equal to the twisting torque minus the torque needed to overcome friction. Hence

accelerating torque $= k_s(\theta_i - \theta_o) - k_f\dot{\theta}_o$

Applying Newton's second law gives

$$J\ddot{\theta}_o = k_s(\theta_i - \theta_o) - k_f\dot{\theta}_o$$

where J is the moment of inertia (second moment of mass) of the structure.
This can be re-arranged as:

$$\ddot{\theta}_o + \frac{k_f}{J}\dot{\theta} + \frac{k_s}{J}\theta_o = \frac{k_s}{J}\theta_i$$

Note again how our modelling assumptions have been carefully chosen not only to be realistic, but also to lead to a tractable mathematical model. A non-linear viscous friction model for example might have been more realistic, but would *not* have led to a system model easy to handle!

This is a *second-order linear differential equation* relating system output θ_o to input θ_i.

Just as was the case for the first-order lag, rather than solve differential equations like this for specific inputs, it is usual to rely on standard step response curves. So instead of substituting particular numerical values of k_s, k_f and J into the antenna equation, let us examine in general terms how we can use a standard model to characterize second-order system behaviour.

The standard form of the second-order equation is written

$$\ddot{\theta}_o + 2\zeta\omega_n\dot{\theta}_o + \omega_n^2\theta_o = k\omega_n^2\theta_i$$

An equation of this type can model many lumped parameter systems which possess inertia, stiffness, and viscous friction. In mechanical terms such models are often known as 'mass–spring–damper' models.

37

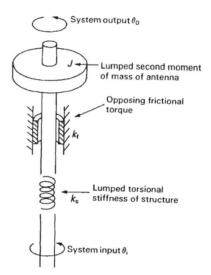

Fig. 3.20   A lumped parameter model of a telescope antenna.

Note that two independent parameters are required for a first-order lag, and three for a second-order lag. The significance of the parameters ζ and ω_n will be considered below.

where ω_n is a constant known as the *undamped natural angular frequency* of the system, k is the steady-state gain, and ζ is a constant known as the *damping factor* or *damping ratio*. The standard parameters ζ, ω_n and k are particularly useful for characterizing second-order system behaviour, just as τ and k did in the first-order case.

Worked Example 3.9   The differential equation model of a telescope antenna structure is:

$$\ddot{\theta}_o + 4\dot{\theta}_o + 100\theta_o = 100\theta_i$$

What is the undamped natural angular frequency and damping factor? What is the significance of the value of the steady-state gain?

Solution   Comparing the system differential equation with the standard form we see that ω_n^2 is equal to the coefficient of the output variable, θ_o. Hence $\omega_n = 10$, and from the coefficient of θ_i, $k = 1$.
   From the coefficient of $\dot{\theta}_o$ we have

$$2\zeta\omega_n = 4$$
$$\zeta = 0.2$$

The significance of a steady-state gain of 1 is that in the steady state a constant θ_i will give rise to a constant θ_o with $\theta_o = \theta_i$. This is exactly what would be expected, since θ_i is the angular displacement of the drive shaft and θ_o the displacement of the antenna structure connected to it. When the antenna is stationary with no external disturbances, these two angles will be the same.

38

Second-order step response

The nature of the step response of a second-order lag varies considerably depending on the value of the damping ratio. (The damping ratio is defined in the way it is in order to characterize these various possibilities.) Normalized responses to a step input are shown in Fig. 3.21.

Note that the output variable is again normalized to the steady-state value k, and the horizontal axis is labelled in terms of normalized time $\omega_n t$. Systems with $\zeta < 1$ are known as *underdamped*, and are characterized by a step response with oscillatory components. The lower the damping, the more pronounced the oscillation, and the closer the observed frequency of oscillation is to ω_n. In fact it can be shown that these damped oscillations occur at a frequency $\omega_{osc} = \omega_n\sqrt{1 - \zeta^2}$. The step response of an *overdamped* system with $\zeta > 1$, on the other hand, does not include any oscillatory components, and rises to its final value without overshoot. A system with $\zeta = 1$ is said to be *critically damped*, and gives the fastest rise in step response without overshoot. A system with $\zeta = 0$ would exhibit sustained oscillations at the undamped natural frequency ω_n (hence the name), and can often be thought of as representing a system on the borderline of instability.

As in the first-order case, these standard responses can be derived by solving the differential equation for various values of ζ, although control engineers are rarely called upon to do so! The precise nature of such time-domain responses will be considered further in Chapter 5; for the time being we shall concentrate on describing, rather than deriving, the important features of the model.

Stability will be discussed in detail in forthcoming chapters.

Worked Example 3.10

An accelerometer has a natural undamped frequency of $100\,\mathrm{rad\,s^{-1}}$ and a damping factor of 0.6. Using the normalized curves of Fig. 3.21 estimate the percentage peak overshoot of the transducer output in response to a step change in

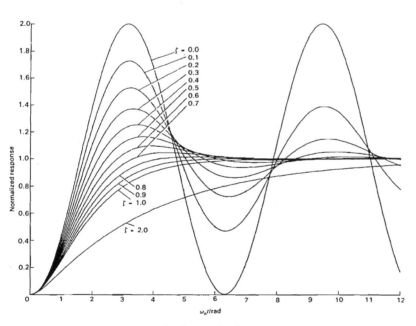

Fig. 3.21 Normalized second-order step response curve.

acceleration. Also estimate how long the output will take to settle within 5% of its final, steady-state value.

Solution From the curves the peak overshoot for $\zeta = 0.6$ is about 10%. The step response curve for $\zeta = 0.6$ suggests that the output will remain within 5% of its steady-state value for values of normalized time greater than $\omega_n t \simeq 5$. Hence settling time $t_s \simeq 5/100$ s or about 50 ms.

Second-order frequency response

The frequency response function of the standard second-order system can again be derived from the differential equation model by transforming into a phasor equation. Since differentiation corresponds to multiplication by the phasor operator $j\omega$, taking the second derivative corresponds to multiplying the associated phasor by $(j\omega)^2$. Hence the differential equation

$$\ddot{\theta}_o + 2\zeta\omega_n\dot{\theta}_o + \omega_n^2\theta_o = k\omega_n^2\theta_i$$

becomes the phasor equation

$$(j\omega)^2\theta_o + 2\zeta\omega_n(j\omega)\theta_o + \omega_n^2\theta_o = k\omega_n^2\theta_i$$

Rearranging gives

$$G(j\omega) = \frac{\theta_o}{\theta_i} = \frac{k\omega_n^2}{(j\omega)^2 + 2\zeta\omega_n(j\omega) + \omega_n^2}$$

Again, $G(j\omega)$ for a second-order lag can be plotted in normalized form as amplitude ratio $|G|/k$ and phase shift ϕ against normalized frequency on logarithmic axes. The normalized frequency in this case is ω/ω_n.

The standard curves are shown in Fig. 3.22. In the second-order case the important points to remember are:

(1) Like the first-order lag, the phase shift is $0°$ at low frequencies. However, in this case it tends towards $-180°$ for $\omega \gg \omega_n$. It is $-90°$ at ω_n irrespective of the system damping ratio.
(2) The high frequency roll-off of the amplitude ratio is 40 dB per decade. The high frequency asymptote intersects the 0 dB axis at $\omega = \omega_n$.
(3) The nature of the frequency response in the region around ω_n is strongly dependent on ζ. In fact, it can be shown that a resonance peak exists for $\zeta < 1/\sqrt{2} \simeq 0.71$ at a frequency $\omega_p = \omega_n\sqrt{1 - 2\zeta^2}$.

As was pointed out earlier, the time-domain and frequency-domain models are absolutely equivalent. This means that the characteristics of one model can be used to deduce those of the other – a very useful technique in control engineering. The following example illustrates this.

Worked Example 3.11   Frequency response tests on a system modelled as second order reveal an amplitude ratio with a resonant peak of 5 dB above the low-frequency level, occurring at a frequency of 4.5 Hz. What will be the expected percentage peak

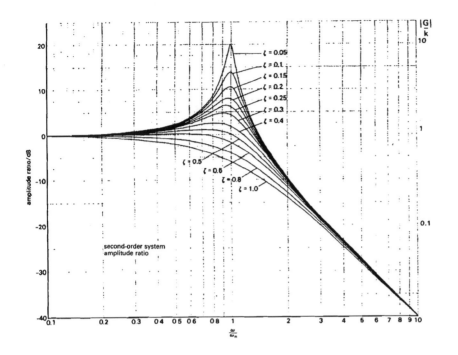

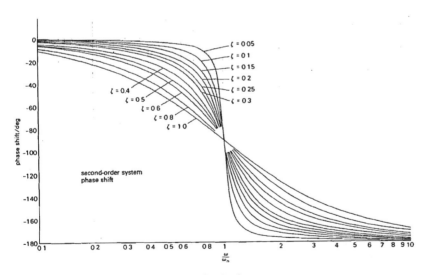

Fig. 3.22 Normalized second-order frequency response curves.

overshoot of the step response and how long after the application of an input step will it occur?

Solution From the standard frequency response curves, a resonant peak of 5 dB corresponds to $\zeta = 0.3$, and occurs at $\omega_p \simeq 0.9\omega_n$ ($\omega_p = \omega_n\sqrt{1 - 2\zeta^2}$). The peak overshoot from Fig. 3.21 is therefore around 35%, and occurs at normalized time $\omega_n t = 3.3$.

The natural frequency of the system is $(4.5/0.9) = 5.0$ Hz. Hence $\omega_n = 10\pi$ and the peak response occurs at time $t = 3.3/10\pi \simeq 0.1$ s.

Higher-order models

The methods developed in the previous sections can be applied to higher-order systems – that is, to systems which are modelled by higher-order linear differential equations. Such equations can be transformed into frequency response functions by repeated use of the phasor differentiation operator $j\omega$. That is,

$$\frac{d}{dt} \to j\omega$$

$$\frac{d^2}{dt^2} \to (j\omega)^2$$

$$\frac{d^n}{dt^n} \to (j\omega)^n$$

and so on. A formal statement of this is as follows. Any nth-order linear differential equation relating system input $x(t)$ to output $y(t)$ can be written:

$$a_n\frac{d^n y}{dt^n} + a_{n-1}\frac{d^{n-1}y}{dt^{n-1}} + \ldots a_1\frac{dy}{dt} + a_0 y$$

$$= b_m\frac{d^m x}{dt^m} + b_{m-1}\frac{d^{m-1}x}{dt^{m-1}} + \ldots b_1\frac{dx}{dt} + b_0 x$$

where $a_0 \ldots a_n$, $b_0 \ldots b_m$ are constant coefficients, $a_n \neq 0$, and $n > m$. This can be transformed into the frequency response function

$$G(j\omega) = \frac{Y}{X} = \frac{b_m(j\omega)^m + b_{m-1}(j\omega)^{m-1} + \ldots b_1 j\omega + b_0}{a_n(j\omega)^n + a_{n-1}(j\omega)^{n-1} + \ldots a_1 j\omega + a_0}$$

The idea of independent, non-interacting subsystems is very similar to systems-oriented electronics design, where input and output impedances of individual subsystems are often designed to prevent one subsystem from loading another.

Fortunately in practice we rarely need anything so complex. Many practical systems can be designed using the models already discussed in this chapter, or similar ones no more complicated. This is because we tend to analyse complex systems in terms of lower-order subsystems. Providing that the subsystems can be treated as essentially independent units, there is a particularly simple way of developing higher-order frequency response models.

As an elementary example, suppose that we wish to use a d.c. servo-motor for position control, as shown in Fig. 3.23(a). In this case output angular displacement is measured by a position transducer, and a suitable error signal is generated by comparing desired and actual position. This time an input–output frequency

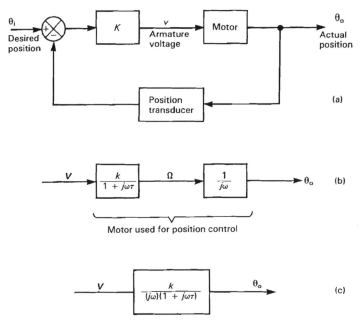

Fig. 3.23 A frequency response model of a position control system.

response model of the motor would relate input armature voltage and output angular displacement. To obtain such a model we could derive a new differential equation, and hence an appropriate frequency response function. However, an alternative approach is to view the new motor subsystem as shown in the block diagram of Fig. 3.23(b). Recall that output velocity is transformed into output position by the operation of integration with respect to time. This fact is represented by the separate integrator block, even though no separate system component is involved. This model is completely equivalent to the single block of Fig. 3.23(c), and the overall frequency response function can be obtained by multiplying together the individual frequency responses.

This is where the logarithmic Bode plot reveals its true value. In terms of Fig. 3.23(b), the frequency response of the first element is

$$G_1(j\omega) = \frac{k}{1 + j\omega\tau}$$

and the frequency response of the integrator is

$$G_2(j\omega) = \frac{1}{j\omega}$$

The frequency response of the overall motor subsystem

$$\frac{\theta_o}{V} = G_3(j\omega) = \frac{k}{(j\omega)(1 + j\omega\tau)}$$

Note that a different model is being used because of the new application. There is no 'single' model associated with a particular system, but rather a range of models defined by individual applications.

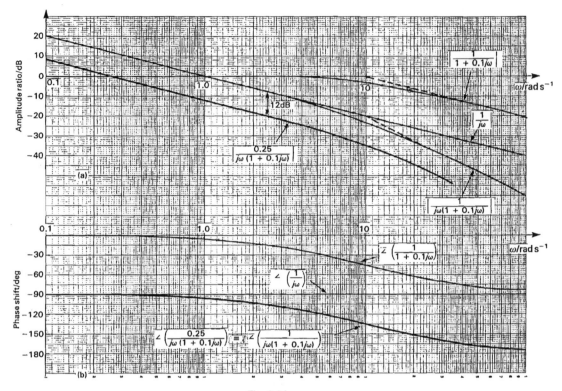

Fig. 3.24

can quickly be sketched by noting that

$$|G_3| = |G_1| \times |G_2|$$

and

$$\phi_3 = \phi_1 + \phi_2$$

This can be extended indefinitely to cascaded components. The rule is to multiply the amplitude ratios and add the phase shifts, no matter how many individual components.

One of the major practical advantages of Bode plots is therefore that the vertical axis of the amplitude ratio is logarithmic, and the vertical axis of the phase shift graph is linear. Multiplication of amplitude ratios corresponds to *adding* decibels, so in both cases we merely add appropriate ordinates when sketching frequency responses of cascaded elements. Furthermore, the straight-line asymptotes resulting from the logarithmic frequency scales make sketching approximate frequency responses particularly easy.

Worked Example 3.12   Sketch the Bode plots of the systems with frequency transfer functions

(a) $G_1(j\omega) = \dfrac{0.25}{(j\omega)(1 + 0.1j\omega)}$

(b) $G_2(j\omega) = \dfrac{1}{(1 + 10j\omega)^2}$

Solution

(a) The time constant of the first-order term is 0.1 s, corresponding to a corner frequency of $10 \, \text{rad} \, \text{s}^{-1}$. Bode plots of $1/j\omega$ and $1/(1 + 0.1j\omega)$ are as shown in Fig. 3.24. Their combination, $1/(j\omega)(1 + 0.1j\omega)$, is obtained by adding individual dB and phase values. The overall resultant, $0.25/(j\omega)(1 + 0.1j\omega)$, differs from this only in the amplitude ratio, which is 12 dB lower owing to the gain of 0.25.

These days computer packages are used to generate Bode plots. It is still a good idea, however, to learn to appreciate how a higher-order plot relates to its lower-order components.

(b) $G_2(j\omega)$ may be viewed as two cascaded first-order lags, each with a low-frequency gain of 1 and a time constant of 10 s. Its Bode plot is shown in Fig. 3.25. Note the high-frequency roll-off of 40 dB per decade, and the amplitude ratio -6 dB at the 'corner frequency' of $0.1 \, \text{rad} \, \text{s}^{-1}$. The phase shift is $-90°$ at the corner frequency and tends to $-180°$ as ω increases indefinitely.

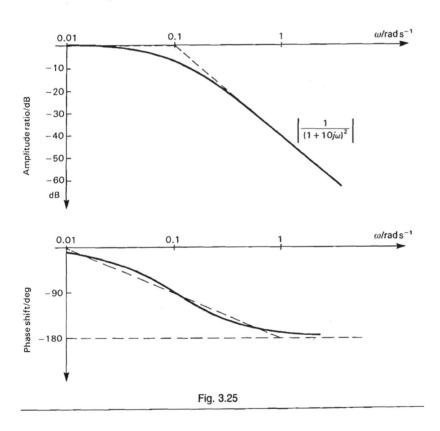

Fig. 3.25

Time delays

The final frequency domain model I want to look at in this chapter is that of a pure *time delay* or *dead time*. Many industrial processes exhibit a unit step response of the form illustrated in Fig. 3.26, which can often be adequately modelled as a first-order lag plus a time delay. We already know how to model the first-order lag in the frequency domain, but what about dead time? Again we look for an appropriate phasor operator, Clearly its magnitude will be unity, since a pure delay does not affect the amplitude of an input sinusoid, but a delay of T seconds introduces a phase lag of ωT radians. The angle of the phasor operator is therefore $\phi(\omega) = -\omega T$ and the complete delay operator is represented by the complex expression $e^{-j\omega T}$. The frequency transfer function corresponding to the broken curve of Fig. 3.26 is therefore obtained by multiplying together the frequency responses of the dead time and the first-order lag:

$$G(j\omega) = \frac{ke^{-j\omega T}}{(1 + j\omega\tau)}$$

A dead time corresponds to a Bode plot whose phase lag increases indefinitely with frequency, rather than levelling out to a constant value at sufficiently high frequencies. The phase characteristic of a pure delay is linear when plotted on a linear frequency axis ($\phi(\omega) = -\omega T$), but there is no simple approximation on the Bode plot with its logarithmic frequency scale.

System analysis and system identification

Earlier in this chapter it was pointed out that system modelling may involve the analysis of the physical behaviour of individual components, or experimental testing, or a combination of both. The *system analysis* route to a mathematical model was illustrated in the derivation of differential equation models of the d.c. motor and the radio telescope, while some features of the experimental approach were indicated in Worked Examples 3.4 and 3.11. Precisely which approach to adopt depends on circumstances. Sometimes the physical processes involved in a system are so complex that the experimental approach is the only feasible way of deriving a model. In other cases it may not be possible to run experimental tests – the components may still be at the design stage, for example, or interruptions to the normal operations of an existing plant may not be permitted.

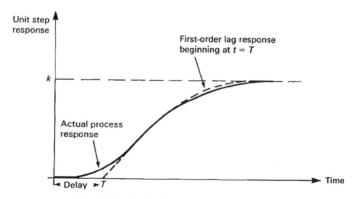

Fig. 3.26　A typical process step response.

The experimental approach is often known as *system identification*. The general technique runs as follows. A system or component is subjected to testing – often a step or frequency response test – and the important features of the recorded data are compared with those of appropriate standard models. If a sufficiently good fit is obtained – and again, the control engineer needs to decide what constitutes a sufficiently good fit on the basis of experience – then the experimental data can be used to estimate values of parameters such as time constant, steady-state gain, damping factor, and so on. This approach treats the system or component being modelled as a 'black box': no knowledge whatsoever of internal behaviour is needed to obtain a model in terms of standard system parameters.

The system analysis route, on the other hand, involves the derivation of a differential equation model by applying physical laws to the internal components of a system, and making appropriate modelling assumptions. In the examples presented in this chapter, for example, we chose to ignore friction in the case of the motor, but to include it in the case of the radio telescope. Again, experience and physical intuition are valuable aids to the designer in taking this sort of decision. Once a differential equation model is obtained, estimates then have to be made of physical parameters such as inertia, friction, viscosity, and so on. In general, some such physical parameters can be predicted or measured accurately, while rough estimates only can be made of others.

It is common to carry out both system analysis and experimental testing, and a comparison of the results obtained is often most illuminating, even when discrepancies exist. Consider the radio telescope, for example. It would not be surprising if values of ζ and ω_n derived by substituting estimates of friction, stiffness and inertia into the differential equations differed considerably from values obtained from step response tests. Physical parameters such as friction and stiffness coefficients are often difficult to estimate accurately, and inaccuracies in the estimates will be reflected in the values deduced for system parameters. Physical analysis still provides much valuable information about the system, however. Unlike identification, analysis allows the designer to predict the effect that changing such *physical* parameters will have on *system* parameters such as natural frequency and damping. Information on the sensitivity of the model to such changes can be obtained even if the values of the physical parameters themselves are not known accurately. Ideally perhaps, both system analysis and testing should be carried out, and the results compared to obtain a 'best fit' model.

Summary

Control engineers most commonly use *linear models* of system elements wherever possible, since these are both sophisticated enough to model a wide variety of physical systems, and simple enough to manipulate and interpret fairly easily. By a *linear model* in this sense is meant a linear differential equation model with constant coefficients. Such models possess the property of 'frequency preservation' – that is, a steady input sinusoid gives rise to a steady output sinusoid at the same frequency, but in general with different amplitude and phase. This property leads to the alternative, frequency-domain, representation of a linear system by means of a frequency transfer function, $G(j\omega)$, where $|G|$ is the amplitude ratio, and

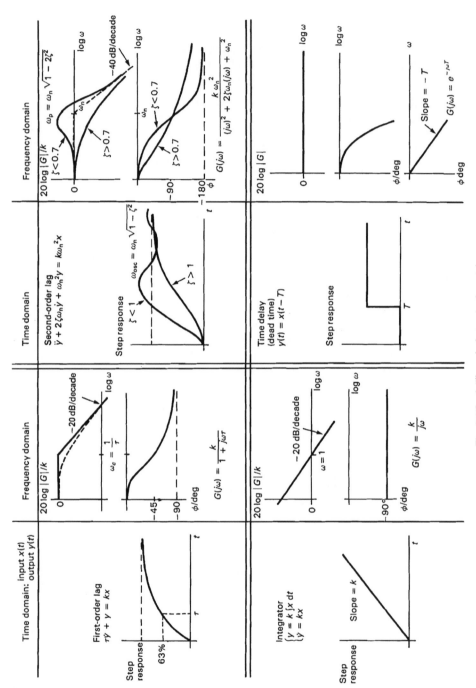

Fig. 3.27 A 'library' of standard models.

$\angle G$ the phase shift of any output sinusoid with respect to the input. Note that we use this alternative representation *not* because the frequency responses of the system components are of particular interest in themselves, but because they lead to useful *models* for system analysis, specification and design.

In this chapter a number of standard models were presented in three alternative guises: (i) differential equations relating input and output in terms of system parameters $(\tau, \omega_n, \zeta, k)$; (ii) time domain step responses; and (iii) frequency response curves. Bode plots were introduced as a convenient representation of the latter. Such standard models offer a common conceptual framework for modelling systems of widely differing physical natures, and are particularly useful when deriving a model from experimental data. The models introduced so far are summarized in Fig. 3.27. They will become much more familiar in the following chapters.

Problems

3.1 Rewrite the following system differential equations in their engineering standard forms. In each case identify values of steady-state gain, time constant, undamped natural angular frequency, and damping ratio as appropriate.
   (a) $2.5\dot{\theta}_o + 2\theta_o = 3\theta_i$ (input θ_i, output θ_o)
   (b) $5\Omega + \dot{\Omega} = v$ (input v, output Ω)
   (c) $5\ddot{\theta}_o + 7\dot{\theta}_o + 20\theta_o = 40\theta_i$ (input θ_i, output θ_o)

3.2 Sketch
   (a) the unit step responses and
   (b) the approximate frequency responses (Bode plots) of the systems represented by the differential equations in Problem 3.1.

3.3 Draw the Bode plots corresponding to the following frequency response functions:
   (a) $100/\{j\omega[(j\omega)^2 + 6j\omega + 100]\}$
   (b) $8/(1 + 40j\omega)^4$

3.4 Frequency response tests on a system modelled as a second-order lag reveal an amplitude ratio with a peak 8 dB above the low-frequency level. The peak occurs at a frequency of 15 Hz. Estimate:
   (a) the damping ratio and natural frequency of the system;
   (b) the percentage peak overshoot of the step response of the system; and
   (c) the time taken for the step response to rise from 10% to 90% of its final value.

4 The frequency response approach to control system design

Objectives
- ☐ To show how open-loop and closed-loop frequency response are related.
- ☐ To introduce the Nichols chart and its use.
- ☐ To present a frequency response stability criterion.
- ☐ To describe integral action and its effects on static error.
- ☐ To outline the steps of a simple frequency response design procedure.
- ☐ To note some ways in which computer tools can aid control system design.

Worked Example 3.4 of the previous chapter illustrated how standard system models can be used for either individual components or complete control systems. In this chapter we shall look at how models of closed-loop systems can be derived from a knowledge of the behaviour of individual system components, and begin to consider how closed-loop systems can be designed to conform to given specifications.

Closing the loop

You should recognize immediately that the motor has a steady-state gain of 0.5 and a time constant of 1 s.

Consider first the simple speed control loop of the type discussed in Chapters 1 and 2, and suppose that in the unity feedback configuration of Fig. 4.1 the motor can be modelled by the transfer function $G(j\omega) = 0.5/(1 + j\omega)$.

Given this information we can immediately derive a closed-loop frequency transfer function for the system, in exactly the same way that an expression for closed-loop *gain* was derived in Chapter 2. We assume that the controller gain, K, does not vary with frequency and that the dynamics of the velocity transducer can be neglected in comparison with those of the motor. (That is, the transducer responds to any changes much faster than does the motor and a unity feedback model can be derived.) Hence we can imagine a frequency response test on the

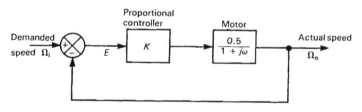

Fig. 4.1   A closed-loop speed control system.

closed-loop system and write, for an input phasor Ω_i (demanded speed), an output phasor Ω_o (actual speed), and an error phasor E,

$$\Omega_o = KG(j\omega)E$$
$$= KG(j\omega)[\Omega_i - \Omega_o]$$

Rearranging gives the *closed-loop frequency response function*:

$$\frac{\Omega_o}{\Omega_i} = \frac{KG(j\omega)}{1 + KG(j\omega)}$$

Note the similarity of this expression to the closed-loop gain expression of Chapter 2. The difference is that $G(j\omega)$ is now a complex quantity with a particular magnitude and phase for each frequency. Hence the term $KG(j\omega)/(1 + KG(j\omega))$ is also complex, and its magnitude and phase at any particular frequency specify the amplitude ratio and phase shift of the *closed-loop* system.

By replacing $G(j\omega)$ by the specific process transfer function $G(j\omega) = 0.5/(1 + j\omega)$, we can investigate the closed-loop behaviour of the motor speed control system in detail.

In this case

$$\frac{\Omega_o}{\Omega_i} = \frac{0.5K/(1 + j\omega)}{1 + 0.5K/(1 + j\omega)}$$
$$= \frac{0.5K}{(1 + 0.5K) + j\omega}$$

Suppose initially that the controller gain $K = 1$. Then we have

$$\frac{\Omega_o}{\Omega_i} = \frac{0.5}{1.5 + j\omega}$$

This is again a first-order lag model. To identify the time constant and steady-state gain, however, it must first be manipulated into standard form $k/(1 + j\omega\tau)$. Dividing numerator and denominator by 1.5 gives

$$\frac{\Omega_o}{\Omega_i} = \frac{0.33}{1 + 0.67j\omega}$$

That is, a closed-loop steady-state gain of 0.33 and a time constant of 0.67 s.

Derive closed-loop frequency transfer functions for the speed control system with (i) $K = 5$ and (ii) $K = 10$. Hence compare the closed-loop step response of the motor speed control system for a controller gain of 1, 5 and 10.

Worked Example 4.1

Solution Proceeding as before we have:

for $K = 5$ $\quad \dfrac{\Omega_o}{\Omega_i} = \dfrac{2.5}{3.5 + j\omega} = \dfrac{0.71}{1 + 0.29j\omega}$

for $K = 10$ $\quad \dfrac{\Omega_o}{\Omega_i} = \dfrac{5}{6 + j\omega} = \dfrac{0.83}{1 + 0.17j\omega}$

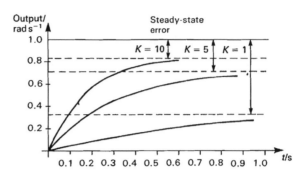

Fig. 4.2 The effect on closed-loop step response of increasing controller gain.

The closed-loop steady-state gains are therefore 0.71 and 0.83, respectively, while the time constants are 0.29 s and 0.17 s. The unit step responses for $K = 1, 5$ and 10 are as shown in Fig. 4.2.

For the general case consider a first-order process with transfer function $G/(1 + j\omega\tau)$, where G is now the steady-state gain. In unity feedback configuration with a proportional controller of gain K the closed-loop transfer function becomes

$$\frac{\Omega_o}{\Omega_i} = \frac{\dfrac{KG}{1 + j\omega\tau}}{1 + \dfrac{KG}{1 + j\omega\tau}} = \frac{KG}{(1 + KG) + j\omega\tau}$$

Dividing numerator and denominator by $1 + KG$ to get the standard form gives

$$\frac{\Omega_o}{\Omega_i} = \frac{\dfrac{KG}{1 + KG}}{1 + \dfrac{1}{(1 + KG)}j\omega\tau}$$

In other words, the unity feedback closed-loop system behaves like a first-order lag, but with a steady-state gain of

$$\frac{KG}{1 + KG}$$

and a new time constant of

$$\frac{1}{(1 + KG)}\tau$$

In theory the time constant and steady-state error of this system will be reduced indefinitely as the controller gain is increased. In practice, of course, the models used to predict the performance will become less adequate as faster response is demanded. As always, mathematical models are only valid within certain limits.

The result of this analysis, and Example 3.1, is entirely consistent with the results obtained in Chapter 2: increasing the controller gain decreases steady-state error (the closed-loop steady-state gain approaches unity). However, increasing the gain has also crucially altered the system dynamics: the time constant becomes progressively shorter as the gain is increased.

Worked Example 4.2

Suppose now that the motor of the previous example is being used to control position, rather than speed, in the way described at the end of Chapter 3. Now a new model of the motor is required, as shown in Fig. 4.3. Describe the salient features of the closed-loop step and frequency response for $K = 1$, 5 and 10.

Solution In this case

$$\frac{\theta_o}{\theta_i} = \frac{KG(j\omega)}{1 + KG(j\omega)} = \frac{0.5K/(j\omega)(1 + j\omega)}{1 + 0.5K/(j\omega)(1 + j\omega)}$$

Multiplying numerator and denominator by $j\omega(1 + j\omega)$ gives:

$$\frac{\theta_o}{\theta_i} = \frac{0.5K}{(j\omega)(1 + j\omega) + 0.5K}$$

This can be rewritten as

$$\frac{\theta_o}{\theta_i} = \frac{0.5K}{(j\omega)^2 + (j\omega) + 0.5K}$$

This transfer function corresponds to a standard second-order model with $\omega_n^2 = 0.5K$ and $2\zeta\omega_n = 1$.

Hence when $K = 1$, $\omega_n = 0.71$ and $\zeta = 0.71$. From the standard second-order response curve for $\zeta = 0.7$ it can be seen that the frequency response is flat at low frequencies, with no resonance peak, while the step response exhibits a 5% overshoot. For $K = 5$, however, $\omega_n^2 = 2.5$, so $\omega_n = 1.58$. Hence $\zeta = 1/3.16 = 0.32$, and the system exhibits a step response with a peak overshoot of about 35%, and a resonance peak in the frequency response of around 5 dB. Increasing the controller gain to 10 gives $\omega_n = 2.24$ and $\zeta \approx 0.22$. Step response peak overshoot is now over 50%, and the frequency response resonance peak more like 8 dB.

Refer to Fig. 3.21 for the standard step-response curves.

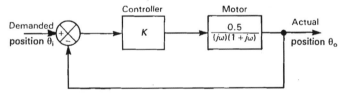

Fig. 4.3 A closed-loop position control system.

In this case, therefore, increasing controller gain has again had the effect of speeding up the response (in the sense that ω_n increases and hence the rise time is decreased), but only at the expense of increasingly oscillatory behaviour, which is likely to be undesirable in any real control system.

You may have noticed that the steady-state gain of the closed-loop system of Fig. 4.3 is unity for all values of K, implying that there is always zero steady-state error for a step change in the desired position. The significance of this will be explored in detail later, but note here that it demonstrates the limitations of the analysis of Chapter 2.

Of course, the previous two examples were chosen so that the closed-loop frequency response corresponded to the standard models of Chapter 3. In practice, though, the situation is usually far more complicated. For example, a particular frequency transfer function which has been used to model the behaviour of steam superheaters in an electricity generating station takes the form

$$G(j\omega) = \frac{K_s e^{-j\omega T}}{(1 + j\omega\tau)^4}$$

where K_s is a constant, τ is of the order of a minute or two, and T is a pure time delay of about 10 s.

In a unity feedback closed-loop system using a proportional controller of gain K, the closed-loop frequency transfer function would take the form

$$\frac{KG(j\omega)}{1 + KG(j\omega)} = \frac{\dfrac{KK_s e^{-j\omega T}}{(1 + j\omega\tau)^4}}{1 + \dfrac{KK_s e^{-j\omega T}}{(1 + j\omega\tau)^4}}$$

Manipulation leads to the closed-loop frequency response expression

$$\frac{KK_s e^{-j\omega T}}{(1 + j\omega\tau)^4 + KK_s e^{-j\omega T}}$$

This is clearly not in any of the standard forms of the previous chapter!

Nevertheless, we can still evaluate the closed-loop frequency response of such systems. So long as some model of the process is available, either as a mathematical expression $G(j\omega)$, or as a set of data points from frequency response tests, the closed-loop transfer function of a unity feedback system can be calculated point by point. In each case the formula $KG(j\omega)/(1 + KG(j\omega))$ can be used.

For each specific value of frequency, $G(j\omega)$ will have a well-defined amplitude and phase (or real and imaginary part). Substituting these values into the closed-loop frequency response expression, and using the normal rules for manipulating complex numbers, will result in values of amplitude and phase (or real and imaginary parts) for the closed-loop frequency response *at the same frequency*. (Note that it is not necessary to know the frequency to perform the arithmetical operations.) The same procedure can be used with amplitude and phase values derived from experiment, even if no analytical model $G(j\omega)$ has been derived. This time, specific experimental values can be substituted into the closed-loop

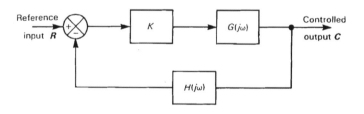

Fig. 4.4 Closed-loop system with feedback path transfer function $H(j\omega)$.

frequency response expression to obtain a pair of closed-loop amplitude and phase values for each open-loop pair.

The same general procedure can be extended to non-unity feedback systems of the form shown in Fig. 4.4. An analysis entirely analogous to that for steady-state closed-loop gain in Chapter 2 leads to a closed-loop transfer function

$$\frac{C}{R} = \frac{KG(j\omega)}{1 + KG(j\omega)H(j\omega)}$$

Check that you can derive this without reference to Chapter 2.

The term $KG(j\omega)$ is known as the *forward path transfer function*, and $KG(j\omega)H(j\omega)$ is known as the *open-loop transfer function* or *open-loop frequency response function*. Note the general form of the closed-loop frequency response as

$$\frac{\text{'forward path'}}{1 + \text{'open-loop'}}$$

As was the case for unity feedback, provided $G(j\omega)$ and $H(j\omega)$ are known, either from an analytical model or from experimental data, then the closed-loop frequency response can be calculated.

The Nichols chart

A particularly convenient way of evaluating unity feedback closed-loop amplitude and phase, given corresponding open-loop values, is provided by the Nichols chart, a special graph paper available commercially and illustrated in Fig. 4.5.

Nichols was another of the American pioneers of feedback control, and a contemporary of Bode. He suggested the form of the Nichols chart in 1947.

The open-loop amplitude and phase – that is, the values corresponding to $KG(j\omega)$ in the convention we have adopted for unity feedback systems – are located using the *rectangular* grid on the Nichols chart. For each pair of values the corresponding closed-loop figures – that is, the amplitude and phase of $KG(j\omega)/[1 + KG(j\omega)]$ – can be read from the *curved* contours.

For unity feedback systems, forward path and open-loop transfer functions are identical – in our convention $KG(j\omega)$.

For example, the curve of Fig. 4.5 was plotted from the following table of open-loop frequency response values.

Frequency ω/rad s^{-1}	Open-loop amplitude ratio/dB	Open-loop phase shift/degrees
0.1	30	-140
0.2	14	-125
0.5	3	-120
1.0	-6	-150
2.0	-9	-174

These open-loop data points were located on the Nichols chart, using the *rectangular grid*, as shown. The closed-loop values can then be read directly using the *curved contours*.

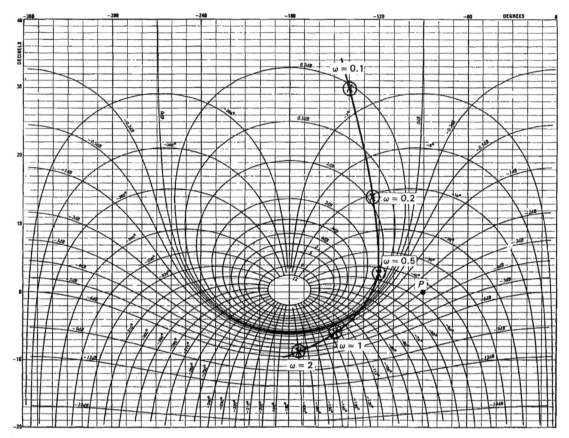

Fig. 4.5 Frequency response data plotted on a Nichols chart.

$\omega/\text{rad s}^{-1}$	Closed-loop amplitude ratio/dB	Closed-loop phase shift/degrees
0.1	0.2	<-2
0.2	1	-10
0.5	1	-45
1.0	-2	-125
2.0	-5	-170

Worked Example 4.3   At a particular frequency the open-loop amplitude ratio and phase shift of a unity feedback system are 1 (0 dB) and $-90°$, respectively. Using the Nichols chart of Fig. 4.5, write down the closed-loop amplitude ratio (in dB) and phase. Check algebraically that this is the result you would expect.

Solution   The 0 dB, −90° point can be located as point P on the rectangular grid of Fig. 4.5. The closed-loop values can then be read from the curved grid as −3 dB, −45°. To check the result algebraically, denote the open-loop transfer function by $KG(j\omega)$. The unity feedback transfer function is hence

$$\frac{KG(j\omega)}{1 + KG(j\omega)}$$

We have $KG(j\omega) = 1 \angle -90° = -j1$.

Hence at the frequency of interest (which we do not need to know), the closed-loop transfer function reduces to the expression $-j1/(1 - j1)$. That is,

$$\frac{1 \angle -90°}{\sqrt{2} \angle -45°}$$

Hence the closed-loop amplitude and phase are $1/\sqrt{2}$ (or −3 dB), −45° as predicted from the Nichols chart.

It is easy to program a computer to carry out the calculation of closed-loop frequency response from open-loop data. However, the Nichols chart still remains an extremely useful tool for design – and also a particularly concise way to present a great deal of information about closed-loop behaviour. This can best be appreciated in the context of an example.

Using the Nichols chart

Figure 4.6 shows the block diagram of a small-scale temperature control system. Step response tests indicate that the process can be modelled as a first-order lag with a time constant of 10 s and a steady-state gain of 1, together with a pure delay (dead time) of 1 s.

What sort of closed-loop behaviour can we expect?

The general technique for analysing closed-loop behaviour using frequency domain modelling is as follows:

(1) Obtain a Bode plot of the open-loop frequency response, using the techniques described in the previous chapter or a computer program if available.
(2) Transfer the open-loop frequency response plot to the Nichols chart, point by point, using the rectangular grid. (Again, commercial computer software is available which displays frequency response data as a Nichols plot.)

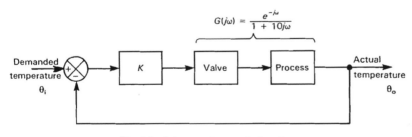

$$G(j\omega) = \frac{e^{-j\omega}}{1 + 10j\omega}$$

Fig. 4.6   A temperature control system.

57

(3) Use the curved grid to determine the characteristics of the closed-loop frequency response.

To apply this procedure to the temperature control system of Fig. 4.6 we need first to choose an appropriate trial value of controller gain K. Now, the steady-state error in response to a step change in demanded temperature can be determined from the closed-loop static gain $KG_{ss}/(1 + KG_{ss})$ where in this case $G_{ss} = 1$. So choosing K around 10 puts the steady-state error in the region of 10%. (We would hope to do better, but this seems to be a suitable starting point.)

In this case, then, the open-loop transfer function is

$$\frac{\theta_o}{\theta_i} = \frac{10e^{-j\omega}}{1 + 10j\omega}$$

where $e^{-j\omega}$ represents the dead time of 1 s.

The time delay contributes to the phase but not the amplitude plot. Using the normalized first-order lag Bode plot of Chapter 3, and the normalized time delay phase characteristic of Fig. 4.7, the Bode plot of the open-loop

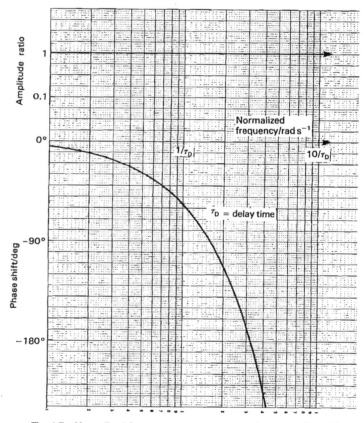

Fig. 4.7 Normalized frequency response curves for a pure time delay.

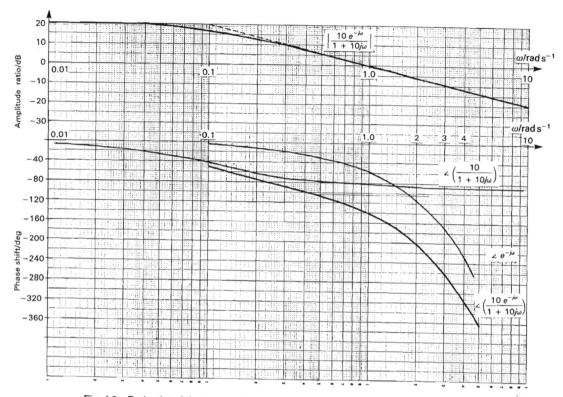

Fig. 4.8 Bode plot of the temperature control system with a proportional controller.

frequency response can be constructed in the usual way. This is illustrated in Fig. 4.8.

Step 2 in the procedure is to transfer the plot point by point to a Nichols chart. It is usual to label each point with the corresponding frequency, as shown in Fig. 4.9. Now the main features of the closed-loop frequency response can be identified. Consider, for example, the amplitude ratio. For low frequencies the amplitude ratio is reasonably flat, but as closed-loop (curved) contours are crossed, it begins to increase, reaching a peak of just about 6 dB (7 dB above the low frequency value of –1 dB) at a frequency of 1.3 rad s^{-1}. In fact, reading from the curved contours the closed-loop amplitude response can be plotted, as in Fig. 4.10.

The closed-loop frequency response enables us to make a prediction of the closed-loop step response. Although the frequency response may not be a close fit to any of our standard models, it seems reasonable to suppose that a resonance peak will correspond to behaviour rather like that of an underdamped second-order system. A 7 dB peak above the low-frequency value would suggest an overshoot of about 50%, in fact. Note, however, that the closed-loop phase (which can also be read from the Nichols chart) has not been taken into account.

The Nichols chart can be visualized as a type of contour map. The curved contours join together points representing equal closed-loop amplitude ratio – just as contours on a map join together points of equal height.

59

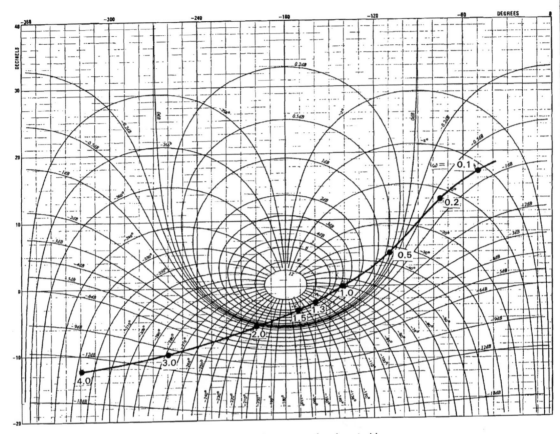

Fig. 4.9 Nichols plot of the proportional control loop.

Note the important *qualitative* difference in closed-loop behaviour between a plant modelled as a first-order lag (e.g. the motor speed control system) and one modelled as a first-order lag plus a time delay (e.g. the temperature control system). The effect of the time delay is to introduce an ever-increasing phase lag as the frequency is increased, tending to push the Nichols plot to the right, viewed in the direction of increasing frequency. Hence, in contrast to the motor example at the beginning of the chapter, the model of the temperature control system predicts instability as gain is increased. Time delays always have such a destabilizing effect on closed-loop behaviour.

We should therefore *not* expect precise correlation between the closed-loop step response and the standard second-order model.

Adjusting the controller gain

Changing the gain K of the proportional controller changes the open-loop amplitude ratio and hence has the effect of shifting a frequency response curve vertically on the Nichols chart. This is illustrated in general terms in Fig. 4.11.

Reducing the gain of the temperature control system, then, has the effect of lowering the curve of Fig. 4.9, and hence reducing the size of the closed-loop resonance peak. Unfortunately, reducing the gain also increases the steady-state error, which is already around 10% for the system under consideration. Increasing the gain to try to reduce this error makes matters worse, as more and more closed-loop contours are crossed. The resonance peak becomes ever higher and the step response more and more oscillatory. In fact, the point at the centre of the

60

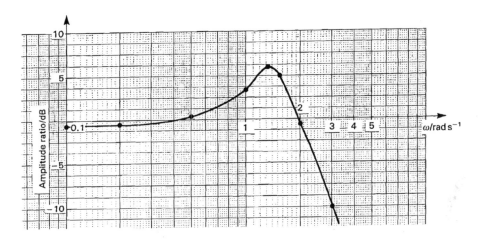

Fig. 4.10 Closed-loop amplitude response under proportional control.

Nichols chart, the $-180°$, $0\,dB$ point on the rectangular, open-loop grid, corresponds to an indefinitely high closed-loop resonance peak. In the terms of a second-order system, the damping ratio becomes zero and the closed-loop system is on the verge of instability since oscillations, once established, never die away.

Worked Example 4.4

What value of controller gain K would drive the temperature control system to the verge of instability?

Solution Raising the frequency response curve by $4\,dB$ causes it to cross the $-180°/0\,dB$ point. $4\,dB$ corresponds to increasing the gain by a factor of F where $20\log_{10} F = 4$. Hence $F = \text{antilog}_{10}(0.2) \simeq 1.6$. The controller gain would then be $10 \times 1.6 = 1.6$.

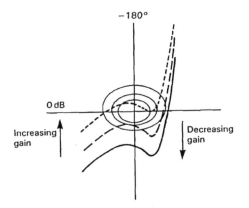

Fig. 4.11 Changing controller gain.

Tracing the curve on a transparent sheet and sliding it up and down a Nichols grid is an easy method of examining the effect of changing the gain of a controller.

61

Stability

The previous discussion is, in fact, an illustration of a general stability criterion, known as the *Nyquist criterion*. In its simplified form, a statement of this criterion is as follows:

> Provided that the open-loop system is stable, then the closed-loop system will also be stable so long as the open-loop $-180°/0$ dB point on the Nichols chart lies to the right of the frequency response curve, viewed in the direction of increasing frequency.

Nyquist published his stability criterion in 1932. Like Bode, he worked on amplifier design, and developed the criterion initially in this context. The Nyquist criterion is not normally stated in quite this form: a more conventional statement is given in Appendix 1.

A rigorous proof of the Nyquist criterion involves the theory of complex variables, and will not be given here. The general closed-loop behaviour discussed above, however, is typical of a wide range of feedback systems. That is, increasing the gain from a low value results in progressively more oscillatory behaviour, culminating in instability.

The analytical form of the closed-loop frequency response function is consistent with such behaviour. At the frequency at which the open-loop ratio is 0 dB and the phase shift is $-180°$, the value of $KG(j\omega)$ is $-1 + j0$. Hence the closed-loop amplitude ratio $|KG(j\omega)/(1 + KG(j\omega))|$ is infinitely great, corresponding to sustained oscillations and the stability borderline. Proper assessment of stability, though, can only be made by assessing the path taken by the curve on the Nichols plot.

Worked Example 4.5

(a) Which of the Nichols plots in Fig. 4.12 represent stable, and which unstable closed-loop systems? Which unstable system(s) can be stabilized by reducing the gain? (The arrows indicate increasing frequency.)

(b) In Fig. 4.12(d) what would be the effect of (i) increasing and (ii) decreasing the gain?

Solution

(a) In Figs 4.12(a) and (b) the $-180°/0$ dB point lies to the left of the frequency response curve, viewed in the direction of increasing frequency. The corresponding closed-loop systems are therefore unstable.

    In (c) and (d) the critical point lies to the right of the frequency response curve, and the closed-loop systems are therefore stable.

    Of the two unstable systems only (b) can be stabilized by reducing the gain. In (a) the $-180°/0$ dB point always lies to the right of the curve, since the open-loop phase shift is greater than $-180°$ for all values of gain.

(b) Increasing the gain corresponds to raising the curve plotted on the Nichols chart. Eventually it will pass to the right of the critical point and the closed-loop system will be unstable. Decreasing the gain at first has a similar effect, but decreasing it sufficiently leads ultimately to a stable closed-loop system again.

Figure 4.12(d) is an example of what is known as a *conditionally stable* system. Examples are not particularly common in practice, but it is interesting to note that it was an electronic (valve) amplifier with similar characteristics which originally prompted Nyquist to investigate the relationship between frequency response and stability.

The closest approach of the frequency response curve of a stable system to the $-180°/0$ dB point on a Nichols chart can be interpreted as a measure of how near the system is to instability. The *relative stability* of a system is often expressed in terms of its *gain* and *phase margins*. These terms are almost self-explanatory, and

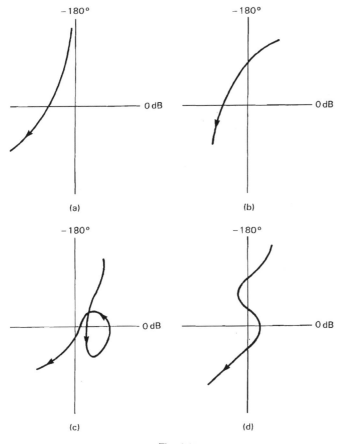

Fig. 4.12

are illustrated in Fig. 4.13. The gain margin is the factor by which the gain must be increased in order to drive the loop into instability (keeping the phase constant), while the phase margin is the additional phase lag necessary to do the same (this time keeping the gain constant). These two parameters are often used to specify control system performance. For example, a particular specification may regard a phase margin of about 45° and a gain margin of 10 dB as producing a satisfactory compromise between relative stability and dynamic performance.

Note that gain and phase margins can be read directly from open-loop Bode plots, although in the absence of a single critical point on the Bode diagram this requires a little more practice. A simple example is shown in Fig. 4.14 – if in doubt the best way is to quickly sketch part of the Nichols plot.

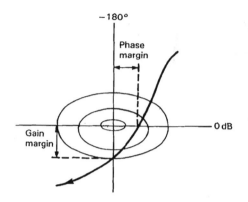

Fig. 4.13 Gain and phase margins on the Nichols chart.

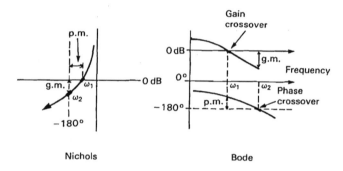

Nichols

Bode

Fig. 4.14 Gain and phase margins on a Bode plot.

Integrating action

The results of the temperature control example force us to re-think our approach to feedback control. The naive analysis of Chapter 2 implied that steady-state error could be progressively reduced, simply by increasing the gain of a proportional controller. The frequency domain discussion of this chapter has shown that problems can arise. Increasing the gain has been shown to reduce steady-state error and to increase the speed of response – but it can also render systems oscillatory and even unstable. Clearly controllers need to be more sophisticated than simple proportional gains.

A common way of reducing steady-state errors is to provide *integral action* in the controller. To see why this can help, consider the motor speed control system again (Fig. 4.1). At some operating point the motor will be running at a constant speed with a particular armature voltage. At this operating point the system is calibrated, so that although distinct, finite input, output and error signals exist, they are all re-defined as equal to zero. When the system is in any other state,

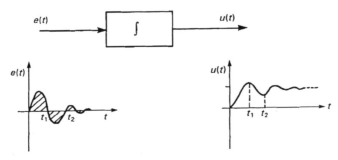

Fig. 4.15 Integral action.

values of the variables Ω_i, Ω_o and e then represent deviations from the calibrated operating point, as described in Chapter 2. Suppose now that the demanded speed is changed to a new, higher value. Clearly, to run the motor at the new higher speed, a greater armature voltage will be required, and this can only be supplied by an increased output from the controller. This increased controller output is equal to Ke, where e is the error signal – that is, the difference between the new demanded speed and the actual speed (all defined as deviations from the operating point). So a proportional controller can never eliminate steady-state error e in this system; it needs the error to be present in order to produce the increased armature voltage.

This problem can be avoided by incorporating integral action in the controller. This means that the time integral of the error is calculated and this value used to provide an output from the controller even when the error is zero, as shown in Fig. 4.15.

While an error exists, the integrator tends to *increase* control action, thus driving the plant towards the demanded output. (Figure 4.15 shows that the integrator output always grows while there is a positive error.) Then, when the error disappears, the continuing integrator output can be used to maintain the control action necessary for the steady-state condition. We can also now understand why in Worked Example 4.2, Fig. 4.3, the steady-state position error was zero for all values of controller gain: the loop includes an integrator *in the process itself*. Arguing again from a physical basis, in the position control system, the controller does not need to generate an increased armature voltage to maintain the new demanded value of output position (assuming the load on the motor is unchanged). The motor will continue to drive the shaft towards the required position for as long as an error signal exists, and then stop. This can be interpreted as the integrating action of the motor itself when used for position control.

While such physical reasoning is an important aid to understanding, a mathematical model of what is happening is also desirable. The mathematical tools necessary for a thorough analysis will be developed in Chapters 5 and 7. However, a frequency domain approach can be adopted which runs as follows. Consider the generalized system of Fig. 4.16. Since the controller is no longer a simple gain, it has been characterized by the frequency transfer function $C(j\omega)$. Normal frequency response analysis leads to the expression:

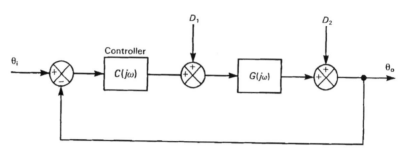

Fig. 4.16 Disturbances in the frequency domain.

$$\theta_o = \frac{C(j\omega)\,G(j\omega)}{1 + C(j\omega)\,G(j\omega)}\theta_i + \frac{G(j\omega)}{1 + C(j\omega)\,G(j\omega)}D_1 + \frac{1}{1 + C(j\omega)\,G(j\omega)}D_2$$

This was one of the major features of an integrator model pointed out in Chapter 3.

Now, if $C(j\omega)$ contains an integrator factor of the form $1/j\omega$, its amplitude ratio will increase indefinitely as the frequency is decreased.

Hence for very slowly varying sinusoidal inputs $C(j\omega)$ will become very large and $C(j\omega)\,G(j\omega) \simeq 1 + C(j\omega)\,G(j\omega)$. The closed-loop frequency response (in the absence of disturbances) therefore approaches unity as the frequency approaches zero, and there is zero steady-state error. By the same argument, the expressions

$$\frac{G(j\omega)}{1 + C(j\omega)\,G(j\omega)} \quad \text{and} \quad \frac{1}{1 + C(j\omega)\,G(j\omega)}$$

tend to zero, indicating that the effects of slowly varying disturbances are also eliminated.

Replace $C(j\omega)$ by K in the above expression.

A corresponding argument holds if the controller is simply proportional with gain K and no integral action, but the process transfer function $G(j\omega)$ contains a factor $1/(j\omega)$. In this case the term $KG(j\omega)/(1 + KG(j\omega))$ approaches unity at low frequencies and the second disturbance term $1/(1 + KG(j\omega))$ tends to zero. Again, the result is zero steady-state error in response to step changes in command input and *demand* disturbances (d_2). Note, however, that this time the first disturbance term $G(j\omega)/(1 + KG(j\omega))$ does *not* tend to zero as the frequency decreases, and steady-state error in response to step changes in *supply* disturbances (d_1) is *not* eliminated. A little reflection should convince you that the integrator must be 'upstream' of the disturbance to remove the static error. Note, however, that the entire preceding argument assumes that the closed-loop system is stable.

This argument is not mathematically rigorous as presented here. A mathematical proof involves a limiting process of complex variables, beyond the scope of this book. Steady-state performance will be discussed in more general terms in Chapter 7.

The proportional + integral controller

Integral action in a controller is usually combined with a proportional gain, such that the controller output, $u(t)$, is the sum of two terms – one proportional to the error, and one proportional to the integral of the error. This is described by the expression

$$u(t) = K\left(e(t) + \frac{1}{T_i}\int e(t)\mathrm{d}t\right)$$

where $e(t)$ is the input error signal and the constant T_i is known as the *integral time*. Commercial controllers offer such P + I (or simply PI) control as one of several options, with a range of permissible values of K and T_i. The designer therefore has two parameters to select, and much greater freedom of action than with proportional or integral control alone.

Write down a frequency response model of a P + I controller, and sketch its Bode plot if $K = 4$ and $T_i = 2\,\text{s}$.

Worked Example 4.6

Solution Adopting the usual notation we have

$$U = K\left(E + \frac{1}{j\omega T_i}E\right)$$

Hence

$$\frac{U}{E} = \frac{K(1 + j\omega T_i)}{j\omega T_i}$$

The frequency response model can be viewed as consisting of the following three terms: a constant gain K, an integrator $1/j\omega T_i$ and the term $(1 + j\omega T_i)$. The latter is simply the inverse of a first-order lag (we could call it a 'lead' term) and its normalized Bode plot is shown in Fig. 4.17. The combined Bode plot for a P + I controller with a gain K of 4 (12 dB) and an integral time of 2 s can be obtained as shown in Fig. 4.18.

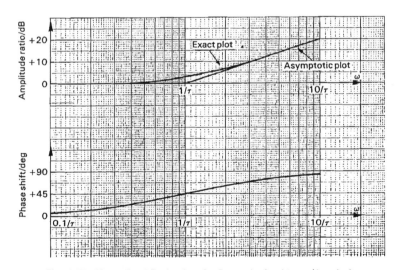

Fig. 4.17 Normalized Bode plot of a first-order lead term $(1 + j\omega\tau)$.

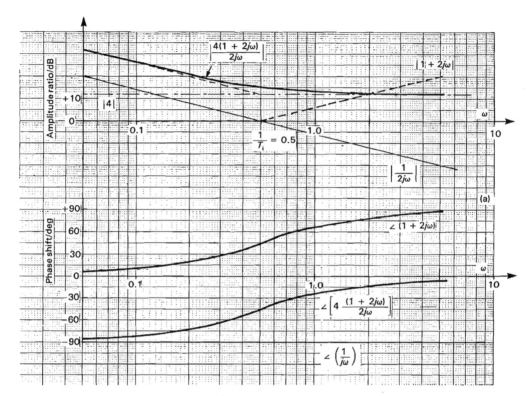

Fig. 4.18

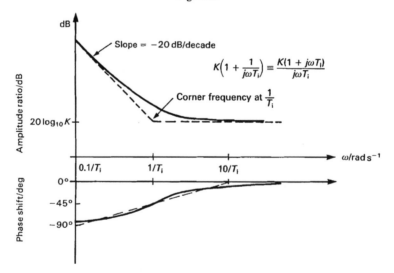

Fig. 4.19 Normalized Bode plot of a P + I controller.

Figure 4.19 shows the Bode plot for a general P + I controller with gain K and integral time T_i. In frequency domain terms, the effect of the integral action in a P + I controller is to boost the gain over a range of 'low' frequencies characterized by the value of T_i. For $\omega \gg 1/T_i$ the controller acts like a proportional controller, since the effect of the integrator is small. The increased low frequency gain for $\omega < 1/T_i$ is accompanied by an additional phase lag, however, and phase lag, as was noted above, can reduce stability margins in a system. This is the price paid for the elimination of steady-state error.

P + I control can be implemented in a number of ways. Until fairly recently pneumatic and analogue electronic devices were most common – and indeed there are still such controllers in use. With the availability of cheap computing power and digital electronics, however, digital controllers have become widespread, either as electronic hardware or as computer software carrying out the P + I calculations digitally. Digital implementation will be considered in Chapters 9–11. Here we concentrate on the broad features of frequency domain controller design.

A design example

The designer's freedom to choose both K and T_i in a P + I controller means that an unsatisfactory open-loop frequency response can often be corrected by the controller with its frequency response of the form of Fig. 4.19. One common approach to design is therefore to shape the open-loop frequency response so as to result in closed-loop behaviour which satisfies given specifications.

Such specifications of closed-loop behaviour are often given in terms of system step response. Typically they might state minimum rise time t_r, maximum settling time t_s, and maximum percentage overshoot, as illustrated in Fig. 4.20. Such specifications imply a trade-off between desirable and undesirable behaviour. For example, we might *require* a given settling time or rise time, and to obtain this speed of response we might *tolerate* an overshoot not exceeding a particular value, if this is unavoidable.

The time domain step response of Fig. 4.20(a) can be translated into frequency response terms as shown in Fig. 4.20(b). Such a response is typical of many practical systems and can be approximated by a standard second-order model.

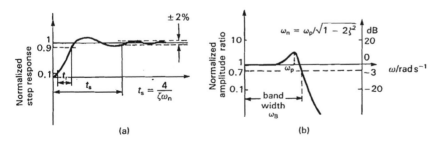

Fig. 4.20 Performance specifications (a) in the time domain and (b) in the frequency domain.

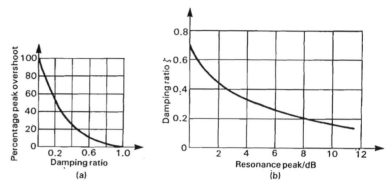

Fig. 4.21 Relationship between damping ratio and (a) peak step response overshoot, (b) amplitude ratio resonance peak.

This approximation allows us to express a specification in terms of standard second-order system parameters, using the graphs of Fig. 4.21 to relate resonance peak, damping ratio and peak overshoot. The 3 dB bandwidth ω_B is also often used as a performance indicator. Broadly speaking, the closed-loop bandwidth gives a measure of the significant frequencies likely to be present in closed-loop system response: in general, the greater the bandwidth the faster the system responds to changes in input, and the faster it counteracts disturbances. Finally, another useful expression is one relating ω_n and ζ to the time taken to settle to within 2% of final value: $t_s \simeq 4/\zeta\omega_n$.

So long as the closed-loop frequency response can be approximated reasonably well by a standard second-order model, it is possible to proceed even further, and relate damping ratio to phase margin, as illustrated in Fig. 4.22. All these graphs should be used with caution, however. The more closed-loop behaviour deviates from ideal second-order behaviour, the less accurate will be the predictions made on the basis of Figs 4.21 and 4.22. Furthermore, estimates of, say, the damping ratio of a system may differ according to whether these are made on the basis of phase margin or closed-loop resonance peak. (In such cases it is good practice to

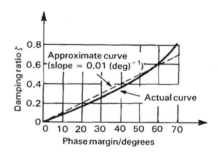

Fig. 4.22 Approximate relationship between phase margin and damping ratio.

assume the 'worst case' estimate for design purposes.) For many commonly occurring systems, however, experience has shown that this general approach is valid, and it is not unusual to find specifications formulated directly in terms of gain and phase margins.

Let us now return to the temperature control system, and examine the effects of using a P + I controller on closed-loop behaviour. From the preceding discussion we would expect the integral term to eliminate steady-state error, but what about dynamic behaviour? Suppose that experience suggests aiming for a phase margin of 45° and a gain margin of 10 dB, with an overshoot of 25% or less.

The new open-loop transfer function of the system with a P + I controller can be written

$$\frac{\theta_o}{\theta_i} = \frac{K}{T_i} \frac{(1 + j\omega T_i)}{j\omega} \frac{e^{-j\omega}}{(1 + 10j\omega)}$$

Our design aim is therefore to choose K and T_i to satisfy gain margin and phase margin specifications. The next few pages will present the stages in such a design,

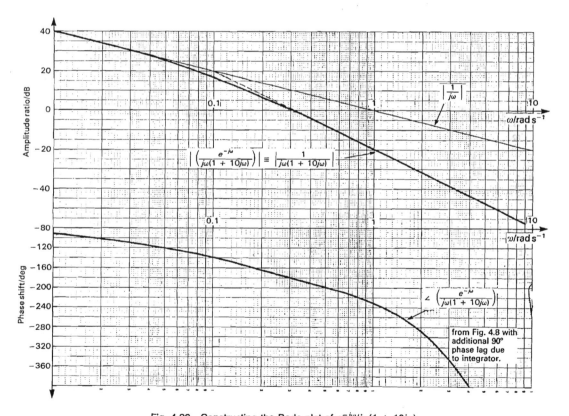

Fig. 4.23 Constructing the Bode plot of $e^{-j\omega}/j\omega(1 + 10j\omega)$.

showing how they can be carried out by hand. The use of various computer tools for such design activities will be discussed in Chapter 12.

Consider first the effect of the $1/j\omega$ term alone, by plotting the frequency response of

This would correspond to a controller consisting of an integrator only, rather than a combination of P + I terms.

$$\frac{1}{j\omega} \frac{e^{-j\omega}}{(1 + 10j\omega)}$$

The corresponding Bode and Nichols plots are shown in Figs 4.23 and 4.24. Both types of plot clearly show that the system is now on the stability borderline with just about zero gain or phase margin. (The integrator has contributed 90° extra phase lag for all frequencies.) However, the effect of reintroducing the $(1 + j\omega T_i)$ term into the numerator of the open-loop transfer function is to reduce this phase lag at higher frequencies. Choosing an appropriate T_i should therefore allow the curve of Fig. 4.24 to be shaped appropriately around the $-180°/0$ dB point. To give a phase margin of 45° we need to reduce the phase lag in this region by around 45–50°.

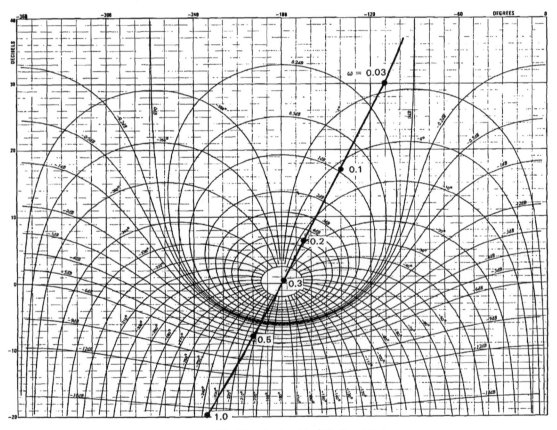

Fig. 4.24   The Nichols plot of $e^{-j\omega}/j\omega(1 + 10j\omega)$.

Determine a value of T_i which will reduce the phase lag by $50°$ at $0.3\,\text{rad}\,\text{s}^{-1}$.   **Worked Example 4.7**

Solution   For a first-order system with a frequency response $1 + j\omega T_i$ the phase is given by

$$\phi = \arctan \omega T_i$$

Hence

$$\frac{\tan 50°}{0.3} = T_i \simeq 4\,\text{s}$$

(Alternatively this can be estimated from the normalized Bode plot of Fig. 4.17.)

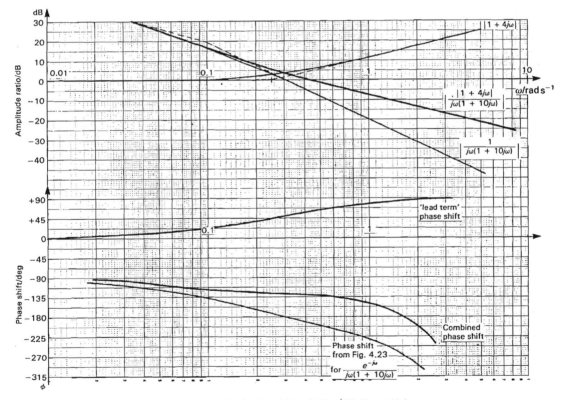

Fig. 4.25   The Bode plot of $(1 + j\omega 4)e^{-j\omega}/j\omega(1 + 10j\omega)$.

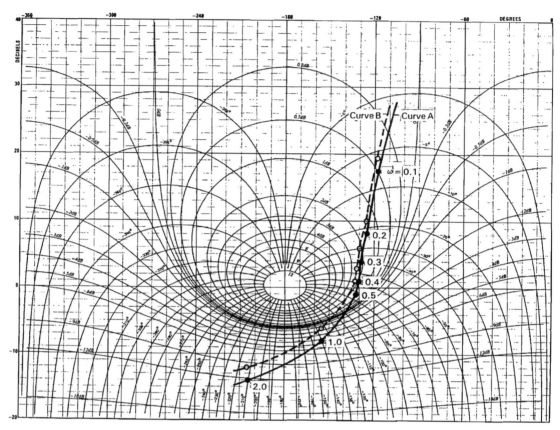

Fig. 4.26 The Nichols plot of $(1 + j\omega 4)e^{-j\omega}/j\omega(1 + 10j\omega)$.

It was tacitly assumed that Fig. 4.24 was a suitable starting point for choosing an appropriate T_i by considering the phase characteristic. In general, however, prior adjustment of the gain may be necessary to locate the response curve in a region of the Nichols chart which will ultimately satisfy specifications. A feel for such design decisions comes with practice.

In general, adjusting K and hence the gain margin affects high frequency behaviour and hence the speed of response. Adjusting T_i will position the phase characteristic. As already noted, however, the two parameters are not independent, and a degree of trial and error (in the light of experience) is likely to be required.

The open-loop transfer function is now

$$\frac{\theta_o}{\theta_i} = \underbrace{\frac{K}{T_i} \frac{(1 + 4j\omega)}{j\omega}}_{\text{controller}} \underbrace{\frac{e^{-j\omega}}{(1 + 10j\omega)}}_{\text{process}}$$

Assuming for the time being that $K/T_i = 1$ (we can make fine adjustments to K later if necessary), the new Bode plot and Nichols plot are shown in Figs 4.25 and 4.26 (bold curve A).

Note that reintroducing the $(1 + 4j\omega)$ term has increased the gain at higher frequencies in comparison with Figs 4.23 and 4.24. The frequency at which the phase lag is $-180°$ (the *phase crossover frequency*) is now substantially higher than the original value of $0.3 \, \text{rad s}^{-1}$.

Figures 4.25 and 4.26 have shown that the phase margin is satisfied with $T_i =$

74

4 s. However, the gain margin, at 12 dB, is a little higher than specified, and may mean that the speed of response may not be sufficient. Raising the whole curve by 2 dB (dotted curve B in Fig. 4.26) reduces the gain margin to 10 dB while continuing to satisfy the phase margin specification.

What value of K is required to satisfy the specification? Estimate the settling time and peak overshoot of the final design.

Solution Curve A of Fig. 4.26 corresponds to $K/T_i = 1$ and hence $K = 4$. Increasing gain by 2 dB corresponds to a factor of 1.26, so the new value of K is about 5. The closed-loop resonance peak for $K = 5$, $T_i = 4$ s occurs at $\omega_p \simeq 0.5$ rad s^{-1}, and has a height of around 2.5 dB. If behaviour similar to a second-order lag is assumed, then $\zeta \simeq 0.4$. Hence, using the expression $\omega_n = \omega_p/\sqrt{1 - 2\zeta^2}$ we have

$$\omega_n \simeq \frac{0.4}{\sqrt{1 - 0.32}} \simeq 0.6$$

The time to settle to within 2% of final value is therefore

$$t_s \simeq 4/\zeta\omega_n = 4/(0.4 \times 0.6) = 17\,\text{s}.$$

Peak overshoot can be estimated from Fig. 4.21(a) as about 25%.

Worked Example 4.8

For an ideal second-order system, ω_n is the frequency at which the closed-loop phase shift is $-90°$. For this system, a closed-loop phase shift of $-90°$ occurs at a frequency of about 0.7 rad s^{-1} for $K = 5$. An alternative estimate of ω_n is therefore 0.7 rad s^{-1}. Since the system is only *approximated* as a second-order system, we should not expect perfect agreement between the two estimates.

Non-unity feedback systems

The preceding design approach relied on the closed loop being analysed in unity feedback form. In many cases, however, transducer dynamics will have to be incorporated, by including an appropriate transfer function $H(j\omega)$ in the feedback path as shown in Fig. 4.27(a). The Nichols chart can still be used, by re-drawing the system as in Fig. 4.27(b) and considering the *transducer* output $b(t)$ as the new system output. The closed-loop frequency response is now

$$\frac{B}{\theta_i} = \frac{KG(j\omega)H(j\omega)}{1 + KG(j\omega)H(j\omega)}$$

and the Nichols plot can be used in the normal way; the open-loop frequency response $KG(j\omega)H(j\omega)$ is plotted on the rectangular grid, and the closed-loop response B/θ_i is read from the curved contours.

The extra $H(j\omega)$ component will often significantly alter closed-loop dynamics. However, it may be possible to specify system behaviour in terms of open-loop gain and phase margins with the transducer included. If, on the other hand, the specific closed-loop response

$$\frac{\theta_o}{\theta_i} = \frac{KG(j\omega)}{1 + KG(j\omega)H(j\omega)}$$

is required it can, of course, be calculated by compensating for the effect of the transducer on B/θ_i as read from the Nichols chart.

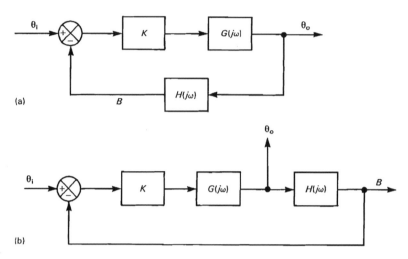

Fig. 4.27 A non-unity feedback system.

$$\frac{\theta_o}{\theta_i} = \frac{B}{\theta_i} \times \frac{1}{H(j\omega)}$$

Whether this final step is actually necessary will depend on the precise form of $H(j\omega)$, and on the performance being required of the closed-loop system.

A note of caution

Like all other design activities, control system design rarely results in a single, 'correct' solution. Rather, any final design is a compromise between competing requirements. The design example in this chapter resulted in values of K and T_i which satisfied gain and phase margin requirements: they may not be the only appropriate values, however, and control engineers need to use their experience and knowledge of the physical realities of the plant to assess a specific design.

To illustrate the sort of consideration involved, let me briefly compare the temperature control example with an alternative approach to the design of a P + I controller for systems with a similar process model. In this alternative approach, the integral time T_i of the controller is set equal to the process time constant. Using the numerical values of the previous example, the open-loop transfer function would become, with $T_i = 10$ instead of 4:

This approach is known as *cancellation*, for obvious reasons. Bear in mind, however, that process parameters are often uncertain and liable to fluctuate, so 'cancellation' can never be exact.

$$\frac{K(1 + 10j\omega)}{10j\omega} \frac{e^{-j\omega}}{(1 + 10j\omega)} = \frac{Ke^{-j\omega}}{10j\omega}$$

The corresponding Nichols plot is shown in Fig. 4.28 for $K = 5$, and at first sight appears to give superior performance, with no resonance peak for the same gain margin, apparently better phase margin, and about the same closed-loop bandwidth.

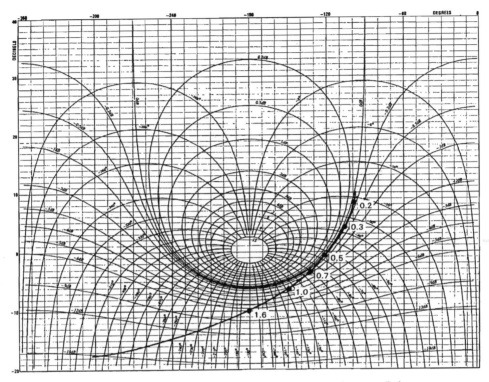

Fig. 4.28 Nichols plot for the temperature control system using cancellation.

To compare the two approaches, however, rather broader criteria are required than the simple specification of bandwidth and relative stability. One important factor not considered so far is *disturbance rejection*, which can often be one of the most important aspects of control system design. This is particularly so for *regulators* – control systems whose major purpose is to maintain the controlled variable constant, at a *set point*, rather than to respond to sudden changes in reference input. In general, the longer the integral time (that is, the *smaller* the integral term in the controller), the longer the effects of disturbances take to die away. By way of illustration, Fig. 4.29 shows the response to unit step changes in input θ_i and disturbance d for a system similar to the temperature control system, but without the time delay. The disturbance rejection properties of the shorter integral time are superior (Fig. 4.29c), although the penalty is overshoot and a longer settling time in the response to a step change in set point (Fig. 4.29b).

Of course, T_i cannot be reduced indefinitely to improve disturbance rejection. Not only might it not be possible to achieve the required stability margins or overshoot specification with a very short integral time, but a smaller T_i also implies a greater control action being applied to the plant. In practical terms, this means more extreme demands on valves and actuators, which may not be able to respond adequately. In any practical applications, such conflicting demands will need to be resolved: control system design always involves such compromises.

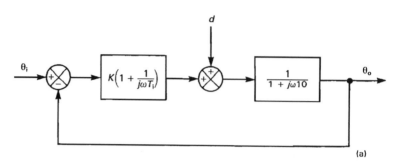

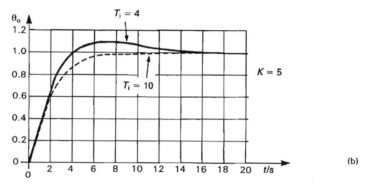

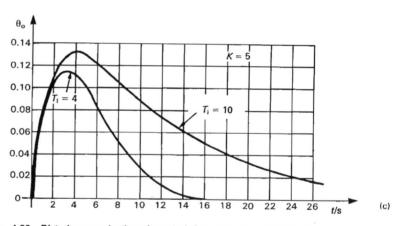

Fig. 4.29 Disturbance rejection characteristics compared. (a) The closed-loop system. (b) Response to a step change in demanded temperature θ_i. (c) Response to a step change in disturbance d.

Summary

The closed-loop frequency response of a system can be evaluated from a knowledge of the frequency transfer functions of the individual components. The *Nichols chart* is a convenient way of doing this for unity feedback systems.

According to the simplified *Nyquist criterion*, a closed-loop system is stable providing:

(a) the open-loop system is stable; and
(b) the open loop $-180°/0\,dB$ point on a Nichols chart lies to the right of the open-loop frequency response curve, viewed in the direction of increasing frequency.

Relative stability is often expressed in terms of the gain and phase margins of a system; these parameters can be determined either from the open-loop Bode plot or the Nichols plot.

Steady-state error in response to a step change in demanded input is removed by the presence of an integrator in the forward path. Similarly, the steady-state error resulting from step changes in disturbance inputs is zero providing there is an integrator upstream of the disturbance in the forward path. A convenient way of introducing an integrator into a control system is to use a P + I controller. The two controller parameters K and T_i can be chosen to produce satisfactory gain and phase margins in many simple systems, but the precise choice in any system involves various performance trade-offs.

Computer-based tools greatly reduce the workload in control system design, but detailed consideration of the various possibilities is postponed to Chapter 12.

Problems

4.1 The following table represents the results of an open-loop frequency response test. Plot the values on a Nichols chart and sketch the closed-loop amplitude ratio. What are the gain and phase margins of the system?

ω	Amplitude ratio/dB	ϕ/deg
0.1	37	-150
0.2	31	-160
0.3	24	-166
0.4	15	-161
0.6	9	-155
1.0	3	-160
1.5	-1	-165
2.0	-4	-170
3.0	-7	-190
5.0	-10	-220

4.2 A liquid level control system is modelled as shown in Fig. 4.30.
(a) Draw the Bode plot of the open-loop frequency response for $K = 1$.

If you have available computer
software of the type discussed in
Chapter 12 it will greatly reduce
the time taken to work through
these problems.

(b) Transfer the Bode plot data to a Nichols chart. Hence estimate the
maximum value of gain K for which you would expect the peak overshoot of the closed-loop step response to be less than 15%.

(c) What are the gain and phase margins for this value of K?

(d) For what value of K would the system be on the verge of instability.

(e) What would be the steady-state error in response to a unit step change
in demanded level, with the value of K derived in (b)?

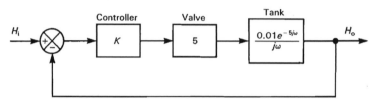

Fig. 4.30

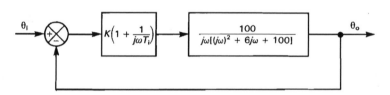

Fig. 4.31

4.3 Figure 4.31 is a simplified model of a position control system for a radio
telescope antenna. To satisfy the various aspects of the specification it is
proposed to use a P + I controller with an integral time of the order of 1 s.

Suggest suitable values of K and T_i to give a gain margin of 6–8 dB, and a
phase margin of 55–60°.

(Note: The Bode plot required in Problem 3.3(a) refers to this system.)

The *s*-plane and transient response   5

□ To introduce the Laplace transform as a technique for modelling signals and   **Objectives**
  systems.
□ To introduce pole–zero diagrams as system models.
□ To explain how the Laplace transform can be used to calculate the complete
  response of a system to an applied input.
□ To distinguish between transient and steady-state components in complete
  system response, and to show how transient response is related to system pole
  positions.
□ To present an *s*-plane stability criterion.
□ To explain the concept of dominant poles, and indicate when a higher-order
  system may be modelled by the dominant pole(s) alone.

The frequency response approach of the previous chapter is quite adequate for designing many simple control systems, and it can be applied to other types of controller in addition to the P and P + I types discussed so far. Later chapters will examine some of these other options available to the control engineer, but first it is useful to develop the system models of Chapter 3 a little further. This will involve presenting a general theoretical framework for modelling the behaviour of linear systems – a theoretical framework, in fact, which is used in many disciplines other than control engineering, albeit in rather different ways (telecommunications and signal processing, for example).

At times, therefore, the material of this chapter may seem rather far removed from practical control systems. However, the models introduced here have proved themselves to be immensely useful for control system analysis (understanding observed behaviour), prediction (suggesting what will happen in new circumstances) and design (causing a system to behave in a desired way). One particularly useful feature of the new models is that they give a convenient representation of system transient behaviour; that is, a mathematical description of how a stable system settles down to its steady-state output after a change in input. This is of great importance in control engineering, since control systems often have to ensure that system variables *change* in an orderly manner – within a given period of time, for example, or without excessive oscillation.

Here I am using the term steady state in its more general sense. For example, we know that if a stable linear system is subjected to a sinusoidal input it will settle down eventually to a steady-state sinusoidal output. At the moment we have no model of how the system behaves between the time the input is switched on and the time the steady state is reached.

The Laplace transform approach

Laplace transform methods are standard modelling tools in control engineering. The general idea is to convert a time domain, differential equation model of system behaviour into a form which is easier to manipulate and interpret. Viewed

I shall simply describe how Laplace transforms are used in control engineering. A more detailed treatment for those who require further information can be found in Meade, M.L. and Dillon, C.R., *Signals and Systems*, Van Nostrand Reinhold, 1986 (2nd edition 1991).

in this light, the use of the *Laplace transformation* in control engineering is very similar to the use of phasor analysis in Chapter 3.

The Laplace transform $F(s)$ of a time domain signal $f(t)$ is defined by the integral

$$F(s) = \int_0^\infty f(t)e^{-st}dt$$

where s is a complex variable, usually known as the Laplace variable, with dimensions 1/[TIME]. There are certain formal mathematical restrictions on the nature of $f(t)$ if $F(s)$ is to exist, but these need not concern us here, since they are satisfied by all practical signals.

Certain characteristics of the transformation do need to be pointed out, however. One of the most important is that the Laplace transform treats a signal as though it were 'switched on' at time $t = 0$ – or, to be more precise, the behaviour of $f(t)$ for $t < 0$ does not affect the value of the integral, owing to the lower limit of $t = 0$. If the integral is evaluated for a particular signal $f(t)$, therefore, the resulting transform $F(s)$ can be used to model a situation where a signal is suddenly applied to the input of a system at time $t = 0$. The significance of this for control engineering purposes is that it makes the Laplace transform ideal for modelling sudden changes to the state of a system: a step change in input, a sudden disturbance, and so on.

In spite of its rather fearsome appearance, and the fact that it transforms a function of a real variable, time, into a function of a complex variable, s, the Laplace transformation is particularly simple to use. Evaluating the Laplace integral for any given signal $f(t)$ defined for $t \geq 0$ leads to a unique corresponding transform $F(s)$.

This equivalence between the two representations is often expressed by the double-headed arrow notation for transform pairs.

$$f(t) \leftrightarrow F(s)$$

Note the convention of lower-case symbols for time domain signals, and upper-case for their Laplace equivalents.

Table 5.1 lists those transform pairs which will concern us in this text, from which it may be seen that the transforms of many commonly occurring signal models are very simple in form. (More extensive tables will be found in the titles given under 'Further Reading'.) Engineers rely on such tables and rarely need to concern themselves with the integral expression itself. Note when using such tables that the form of the Laplace integral implies the following properties:

1. If

$$x(t) \leftrightarrow X(s)$$

and

$$y(t) \leftrightarrow Y(s)$$

then

$$x(t) + y(t) \leftrightarrow X(s) + Y(s)$$

2. If

$$x(t) \leftrightarrow X(s)$$

Table 5.1

$f(t), t > 0$	$F(s)$
1	$\dfrac{1}{s}$
t	$\dfrac{1}{s^2}$
e^{-at}	$\dfrac{1}{(s + a)}$
te^{-at}	$\dfrac{1}{(s + a)^2}$
$\sin \beta t$	$\dfrac{\beta}{s^2 + \beta^2}$
$\cos \beta t$	$\dfrac{s}{s^2 + \beta^2}$
$e^{-at} \sin \beta t$	$\dfrac{\beta}{(s + a)^2 + \beta^2}$
$e^{-at} \cos \beta t$	$\dfrac{(s + a)}{(s + a)^2 + \beta^2}$

and K is a constant, then

$$Kx(t) \leftrightarrow KX(s)$$

In addition, when deriving forms of signals $f(t)$ from transforms $F(s)$, always bear in mind that $f(t)$ is zero before time $t = 0$. Hence, for example, if the transform $F(s)$ of a signal is $1/(s^2 + 1)$, then $f(t) = \sin t$ after time $t = 0$ *and zero before* $t = 0$. Some texts note this explicitly by writing the signal as $f(t) = u(t) \sin t$ where $u(t)$ is the unit step function, equal to 1 after time $t = 0$ and zero before.

These properties are a consequence of the linearity of the transform operation.

Using Table 5.1 and the linearity properties of the Laplace transform, complete the following table of signals and transforms.

Worked Example 5.1

$f(t)$ $(t > 0)$	$F(s)$
(a) $5e^{-2t}$	
(b)	$\dfrac{2}{s^2}$
(c)	$\dfrac{3}{s + 4}$
(d) $e^{-t} + \frac{1}{2}e^{-3t}$	
(e) $1 + \sin 3t$	
(f)	$e^{-t} \sin 2t$

Solution From the table we have

(a) $\dfrac{5}{s + 2}$

(b) $2t$

(c) $3e^{-4t}$

Note in (f) that the transform of $e^{-t}\sin 2t$ is *not* equal to the transforms of e^{-t} and $\sin 2t$ multiplied together. Multiplication (except by a constant) is *not* a linear operation.

(d) $\dfrac{1}{s + 1} + \dfrac{1}{2(s + 3)} = \dfrac{2(s + 3) + s + 1}{2(s + 1)(s + 3)} = \dfrac{3s + 7}{2(s + 1)(s + 3)}$

(e) $\dfrac{1}{s} + \dfrac{3}{s^2 + 9} = \dfrac{s^2 + 3s + 9}{s(s^2 + 9)}$

(f) $\dfrac{2}{(s + 1)^2 + 4} = \dfrac{2}{s^2 + 2s + 5}$

I drew attention above to the notion of using Laplace transforms to convert differential equation models into a more convenient form, rather as phasors were used to obtain frequency response functions. The key to the phasor techniques of Chapter 3 was to transform the time domain operation of differentiation (of a steady sinusoid) into the more convenient phasor operation of multiplication by $j\omega$. A similar, but much more generally applicable, procedure is available for Laplace transforms. Let me first quote the formal mathematical theorem, and then explain why it is nowhere near so formidable as it looks!

If $f(t)$ has the Laplace transform $F(s)$, then the transform of the nth derivative of $f(t)$, $d^n f(t)/dt^n$, is given by the expression

$$s^n F(s) - s^{n-1}f(0) - s^{n-2}f'(0) - \ldots f^{(n-1)}(0)$$

where $f(0)$, $f'(0)$, $\ldots f^{(n-1)}(0)$ are the values of $f(t)$ and its first to $(n - 1)$th derivatives at time $t = 0$.

Now, the first term in this expression is simply $F(s)$ multiplied by s^n. The other terms all relate to the value of $f(t)$ and its derivatives at time $t = 0$, and this fact brings about a great simplification. For the purposes of constructing a generally applicable system model we normally assume that a system is in a *quiescent* state before the application of an input. By a 'quiescent state' is meant that all relevant system variables and their derivatives are zero at time $t = 0$. This assumption allows us to concentrate on the general characteristics of the system, rather than the effect of any specific initial conditions.

In control engineering, all system variables are defined with reference to a fixed operating point, as described in earlier chapters. Hence the quiescence condition can be interpreted as implying that all system variables are at their defined operating levels, unchanging with time.

The general differentiation theorem must still be used whenever we wish to build into our model initial conditions other than zero. A common example is when using Laplace transforms to solve differential equations with specific non-zero initial conditions. Then the quiescence condition at time $t = 0$ does not apply, and the additional terms in the Laplace representation of differentiation take into account the initial conditions of the problem.

Under these circumstances, the Laplace transform representation of differentiation becomes particularly simple. Differentiation of any system variable $f(t)$ corresponds simply to multiplication of the transform $F(s)$ by s; differentiation twice corresponds to multiplication by s^2; and so on.

Armed with this differentiation property of Laplace transforms, it is straightforward to convert a differential equation modelling the behaviour of a linear system into a completely equivalent transform representation. For example, in the case of the familiar first-order differential equation

$$\tau \dot{y} + y = kx$$

where, as usual, x and y represent time-varying input and output, respectively, each term can now be transformed immediately into its Laplace counterpart: $x(t)$ becomes $X(s)$; $y(t)$ becomes $Y(s)$; and $y'(t)$ becomes $sY(s)$. Hence we can write

$$\tau s Y(s) + Y(s) = kX(s)$$

Collecting terms and rearranging gives

$$Y(s) = X(s)[k/(1 + s\tau)]$$

You may have noticed the formal similarity between these manipulations, and the transformation of the same differential equation into a phasor equation and hence a frequency transfer function in Chapter 3. If we let

$$G(s) = k/(1 + s\tau)$$

the above expression becomes

$$Y(s) = X(s) \times G(s)$$

where $G(s)$ is a function of s such that, in general, $G(s)$ takes on a specific complex value for each complex value of s. This expression is an input–output relationship between transforms (assuming the initial quiescent condition), much as the frequency response function $G(j\omega)$ is an input–output relationship between sinusoids (assuming the steady-state condition).

If we write out in full the two expressions

$$G(s) = \frac{k}{1 + s\tau}$$

and

$$G(j\omega) = \frac{k}{1 + j\omega\tau}$$

the close relationship between the two models is even more striking. Specifically, if we substitute any purely imaginary value of s into $G(s)$ ($s = j\omega_0$, say), then we obtain a complex number representing the frequency response of the system at the frequency ω_0. For example, the complex number

$$G(s)|_{s=j2} = G(j2) = \frac{k}{1 + j2\tau}$$

represents the frequency response of the first-order lag at an angular frequency of $\omega = 2$ rad s^{-1}.

In the light of the preceding discussion it should come as no surprise that $G(s)$ is known as a transfer function. Formally we can write

$$G(s) = \frac{Y(s)}{X(s)} = \frac{\text{Laplace transform of system output}}{\text{Laplace transform of system input}}$$

(assuming the quiescence condition before $t = 0$).

In addition to the above differentiation rule for transforms, there is a corresponding rule for integration. Providing the initial quiescence condition holds we can sum up these two rules as:

If

$$f(t) \leftrightarrow F(s)$$

then

$$df/dt \leftrightarrow sF(s)$$

and

$$\int_0^t f \, dt \leftrightarrow \frac{1}{s} F(s)$$

Using these rules, the Laplace transformation converts a time domain equation involving derivatives and integrals into an algebraic equation in the Laplace variable, s. Such algebraic equations can then be manipulated into transfer function form, and used to develop the extremely versatile system models mentioned at the beginning of this chapter.

Worked Example 5.2 Write down the transfer function in s of (a) the standard second-order model represented by the differential equation

$$\ddot{y} + 2\zeta\omega_n\dot{y} + \omega_n^2 y = k\omega_n^2 x \qquad \text{(input } x, \text{ output } y)$$

(b) the P + I controller defined by output

$$u(t) = K\left[e(t) + \frac{1}{T_i} \int e(t) \, dt \right]$$

Solution

(a) Transforming term by term for a general input and output we have

$$s^2 Y(s) + 2\zeta\omega_n s Y(s) + \omega_n^2 Y(s) = k\omega_n^2 X(s)$$

Hence

$$\frac{Y(s)}{X(s)} = \frac{k\omega_n^2}{s^2 + 2\zeta\omega_n s + \omega_n^2}$$

(b) Again we can transform term by term, this time using the above integration rule.

$$U(s) = K\left[E(s) + \frac{E(s)}{sT_i} \right]$$

Hence

$$\frac{U(s)}{E(s)} = K\left(1 + \frac{1}{sT_i}\right) = \frac{K}{T_i}\left(\frac{1 + sT_i}{s}\right)$$

Note again that if s is replaced by $j\omega$, these transfer functions become the corresponding frequency response functions derived in Chapter 4.

Poles and zeros

The transfer function $G(s)$ of a linear system is in general a ratio of two polynomials in s. (The major exceptions in control engineering are systems with dead time, which possess terms of the form e^{-sT} in the transfer function.)

In the general case, a system transfer function can be written as

$$G(s) = \frac{b_m s^m + b_{m-1} s^{m-1} + \dots b_1 s + b_0}{a_n s^n + a_{n-1} s^{n-1} + \dots a_1 s + a_0}$$

where $b_0, b_1 \dots b_m$ and $a_0, a_1 \dots a_n$ are the real coefficients in the differential equation modelling the system, and $m < n$ for a practical system.

In factorized form, then

$$G(s) = K \frac{(s - z_1)(s - z_2) \dots (s - z_m)}{(s - p_1)(s - p_2) \dots (s - p_n)}$$

where z_1 to z_m, and p_1 to p_n, are the roots of the numerator and denominator respectively. Apart from a constant multiplier or gain factor K, therefore, such a $G(s)$ can be completely specified by the roots of the numerator and denominator polynomials, z_1 to z_m and p_1 to p_n. These values of s are known as *zeros* and *poles*. A zero is a value of s which makes the numerator of $G(s)$, and hence $G(s)$ itself zero: a pole is a value which makes the denominator zero, and hence $G(s)$ becomes infinitely great. For example, if $G(s) = (s + 1)/s$, there is a pole at $s = 0$ and a zero at $s = -1$. Note that zeros and poles must either be real or occur in complex conjugate pairs, since they are the roots of polynomials with real coefficients.

The poles and zeros of a transfer function can be represented on a two-dimensional diagram, in the usual way for complex quantities. The real part of the complex variable s is generally denoted σ and the imaginary part $j\omega$, and the complex plane representing all values of $s = \sigma + j\omega$ is known as the *s-plane*.

The significance of the imaginary part $j\omega$ has already been described: substituting $s \to j\omega$ in $G(s)$ gives the frequency response function $G(j\omega)$. The physical interpretation of the real part σ of a particular value of s will be explored later in this chapter.

Figure 5.1 shows the *pole–zero plot* for $G(s) = (s + 1)/s$. It is conventional to mark the position of a pole with a cross and a zero with a small circle. Note that an identical plot would be obtained for $G(s) = K(s + 1)/s$. The constant multiplier K must be specified separately to give a complete description of the transfer function.

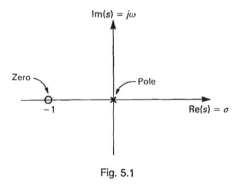

Fig. 5.1

Sketch the pole–zero configurations of:

(a) a P + I controller with an integral time of 2 s;
(b) the second-order system $G(s) = 1/(s^2 + s + 1)$.

Worked Example 5.3

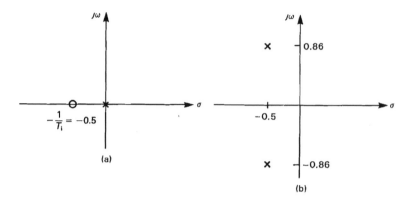

Fig. 5.2

Solution

(a) A P + I controller has the transfer function

$$\frac{U(s)}{E(s)} = \frac{K(1 + T_i s)}{T_i s}$$

Hence there is a pole at $s = 0$ and a zero at $s = -1/T_i$. For $T_i = 2$, therefore, the zero lies at $s = -0.5$.

(b) The denominator of the transfer function

$$G(s) = \frac{1}{s^2 + s + 1}$$

has complex roots. Hence the system possesses a complex conjugate pair of poles where $s^2 + s + 1 = 0$. That is, where

$$s = -\frac{1}{2} \pm j\frac{\sqrt{3}}{2}$$

Since the numerator is a constant, there are no zeros.

Figure 5.2 shows the pole–zero plots. Note again that in (a) the controller gain does not affect the pole–zero diagram.

A pole–zero plot, while expressing all the information contained in a transfer function $G(s)$ (apart from the constant multiplier), does not illustrate explicitly the behaviour of the function other than where it takes the values zero and infinity. More detail about the general behaviour of $G(s)$ can be shown on a three-dimensional diagram, where the magnitude $|G(s)|$ of the transfer function is shown as the height of a surface above the s-plane. Figure 5.3 shows such a three-dimensional plot for $G(s) = (s + 1)/s$. The pole at $s = 0$ can be thought of as 'holding up' the surface, while the zero at $s = -1$ 'pegs it down'.

With practice, it is not difficult to visualize approximately such a 3D surface for

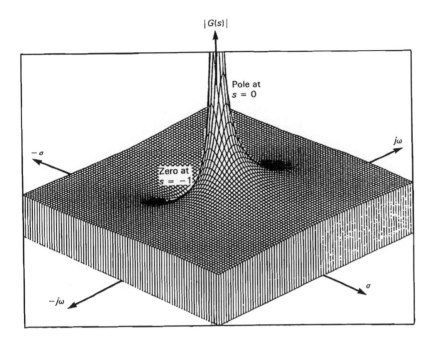

Fig. 5.3   The $|G(s)|$ surface for $G(s) = s/(s + 1)$.

a given pole–zero plot. Once such a surface is visualized, it is then straightforward to obtain a general idea of the amplitude ratio of the frequency response of the system. Recall that the frequency response function $G(j\omega)$ is obtained from $G(s)$ by allowing s to take on imaginary values $s = j\omega$ only. In terms of the $|G(s)|$ surface, then, the amplitude ratio of the frequency response, $|G(j\omega)|$ corresponds to the height of the 'cut' in the 3D surface along the $j\omega$ axis from $\omega = 0$ to $\omega = \infty$.

This is illustrated for a first-order system in Fig. 5.4(a) and for an underdamped second-order system in Fig. 5.4(b). (Note, however, that pole–zero plots normally use linear axes, so the amplitude frequency response curves visualized in this way will not correspond closely to Bode plots on logarithmic axes.)

Calculating system response

Since the transfer function as defined above contains the built-in assumption that the system is quiescent at time $t = 0$, it is easy to model the complete response of a linear system to a sudden change in input. This is highly relevant to understanding how control systems will respond to sudden changes in demanded input or disturbances. The fundamental relationship is

$$G(s) = \frac{Y(s)}{X(s)} \quad \text{or} \quad Y(s) = G(s) \times X(s)$$

A basic knowledge of the technique of partial fractions is assumed in what follows. A detailed exposition can be found in, for example, Kuo, B.C., *Automatic Control Systems*, 5th ed., Prentice-Hall, 1987; or Kuo, F.F., *Network Analysis and Synthesis*, Wiley, 1966.

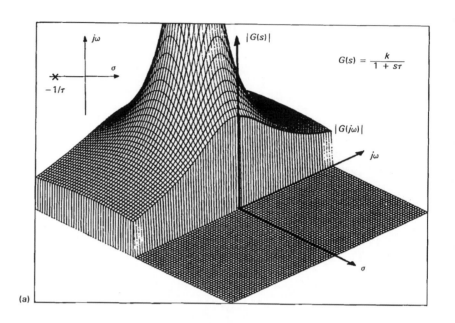

$$G(s) = \frac{k}{1 + s\tau}$$

(a)

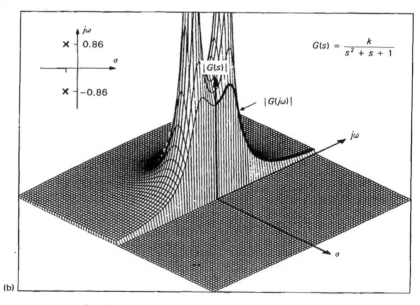

$$G(s) = \frac{k}{s^2 + s + 1}$$

(b)

Fig. 5.4 $|G(s)|$ surfaces for first and (underdamped) second-order systems.

90

introduced in the previous section. The general procedure for calculating system response is to use standard tables to find the transform $X(s)$ of the input, carry out the multiplication of $X(s)$ and the system transfer function $G(s)$, and then use tables again to find the time domain output corresponding to $Y(s)$. The final step will often involve expressing the product $G(s) \times X(s)$ as a series of *partial fractions*.

Worked Example 5.4

(a) Use the partial-fraction method to evaluate the unit step response of a first-order system with the transfer function $G(s) = 1/(s + 2)$.

(b) Find the signal $f(t)$ corresponding to the Laplace transform

$$F(s) = \frac{(s + 1)}{(s + 2)(s + 3)}$$

Solution

(a) Here the transform of the input, $X(s) = 1/s$ for the unit step. Hence

$$Y(s) = \frac{1}{s(s + 2)}$$

Now, let

$$\frac{1}{s(s + 2)} = \frac{A}{s} + \frac{B}{s + 2}$$

where A and B are constants. One way of evaluating A and B is to multiply both sides by $s(s + 2)$, and write the above expression as

$$1 = A(s + 2) + Bs$$

Substituting $s = 0$ eliminates the term in B, giving $A = \frac{1}{2}$.
Similarly, substituting $s = -2$ gives $B = -\frac{1}{2}$.
Hence

$$Y(s) = \frac{1}{2}\left(\frac{1}{s} - \frac{1}{s + 2}\right)$$

Using the table again we have $y(t) = \frac{1}{2}(1 - e^{-2t})$ (for $t > 0$) as the step response.

(b) Proceeding as above we have

$$\frac{(s + 1)}{(s + 2)(s + 3)} = \frac{A}{(s + 2)} + \frac{B}{(s + 3)} = \frac{A(s + 3) + B(s + 2)}{(s + 2)(s + 3)}$$

Hence

$$s + 1 = A(s + 3) + B(s + 2)$$

Substituting

$s = -2$ gives $-1 = A$
$s = -3$ gives $-2 = -B$

Hence

$$F(s) = \frac{2}{(s + 3)} - \frac{1}{(s + 2)}$$

$$f(t) = 2e^{-3t} - e^{-2t} \text{ for } t > 0$$

This is what we would expect by writing the first-order transfer function in standard form as $G(s) = 0.5/(1 + 0.5s)$. The steady-state gain is 0.5 and the time constant is 0.5 s.

Check for yourself that in each case addition of the partial fraction terms gives the original expression.

Step response in the general case

Extending this approach to a general linear system (defined by its transfer function) allows an important link to be made between the system poles and the general nature of the response. In control engineering terms, this means that the pole–zero model of a component or a complete system can be used to give valuable *quantitative*, as well as qualitative, information about how the system behaves. To illustrate this, consider the step response of a system with a transfer function

A similar argument applies for inputs other than a step, so the results of this section are of general significance for system behaviour.

$$G(s) = \frac{N(s)}{D(s)}$$

where the order of the numerator polynomial $N(s)$ is lower than that of the denominator $D(s)$. Suppose further that the roots of $D(s)$ and hence the poles of the transfer function $G(s)$ are real, distinct and non-zero. That is,

$$G(s) = \frac{KN(s)}{(s + a)(s + b)(s + c)\dots}$$

where $a, b, c \dots$ are real and distinct and K is a constant. The transform of the step response of the system, $G(s)/s$ can now be expanded as a series of partial fractions

$$\frac{KN(s)}{s(s + a)(s + b)(s + c)\dots} = \frac{K_0}{s} + \frac{K_1}{(s + a)} + \frac{K_2}{(s + b)} + \frac{K_3}{(s + c)} + \dots$$

This transform can be viewed conveniently as consisting of two parts. The first, the single term K_0/s, is associated with the particular input – in this case a unit step input with the transfer function $1/s$. The remainder consists of the terms

$$\frac{K_1}{(s + a)} + \frac{K_2}{(s + b)} + \frac{K_3}{(s + c)} + \dots$$

associated with the system poles at $s = -a$, $-b$ and $-c$, respectively.

Transforming these latter terms back into the time domain using Table 5.1 gives the following components in the step response:

$$K_1 e^{-at} + K_2 e^{-bt} + K_3 e^{-ct} + \dots$$

That is, each real pole in the system transfer function corresponds to an exponential term in the complete response. The values of the constants K_1, K_2, $K_3 \dots$ are affected by $N(s)$, but the general exponential forms of the individual terms depend only on the pole positions.

Suppose now that in addition to real roots, the denominator $D(s)$ of the transfer function also possesses a number of distinct, complex conjugate pairs of roots. It is most convenient to handle these (with no loss of generality) by assuming quadratic factors in $D(s)$ of the form $[(s + \alpha)^2 + \beta^2]$, corresponding to poles at $s = -\alpha \pm j\beta$. A partial fraction expansion of the step response can again be obtained, but this time each pair of complex poles will give rise to a second-order term of the form:

$$\frac{As + B}{[(s + \alpha)^2 + \beta^2]}$$

in the expansion, where A and B are constants again depending on the particular system and input.

The time domain equivalent of this term does not feature explicitly in Table 5.1. However, we can rewrite the expression so as to use the table of standard forms:

Let

$$\frac{As + B}{(s + \alpha)^2 + \beta^2} = \frac{A(s + \alpha)}{(s + \alpha)^2 + \beta^2} + \frac{C\beta}{(s + \alpha)^2 + \beta^2}$$

where $B = A\alpha + C\beta$.

The time domain equivalent from Table 5.1 is hence

$$Ae^{-\alpha t} \cos \beta t + Ce^{-\alpha t} \sin \beta t$$

which, using the properties of sinusoids, can be combined as the single term

$$Re^{-\alpha t} \sin(\beta t + \phi)$$

Specific values of the new constants R and ϕ can be derived if desired from the original constants A and B in the partial fraction expansion. Here, however, we are more concerned with the general form of the term. The important point to note is that just as a real pole at $s = -a$ corresponds to a component Ke^{-at} in the system response, so a complex conjugate pair of poles at $s = -\alpha \pm j\beta$ corresponds to an oscillatory component $Re^{-\alpha t} \sin(\beta t + \phi)$ in the response. What is more, all such components will eventually die away to zero so long as all terms like $-a$ and $-\alpha$ (the real parts of the pole positions) are negative – that is, *so long as the poles of $G(s)$ lie in the left half of the s-plane.* The further to the left of the $j\omega$-axis the poles lie, the faster the component will die away. This is illustrated in Fig. 5.5, which shows the general forms of time domain response terms associated with various regions of the s-plane. Similarly, the further a pair of complex poles lies from the real, σ-axis, the greater the value of β in the term $Re^{-\alpha t} \sin(\beta t + \phi)$, and the higher the frequency of the oscillatory component in the response.

If, on the other hand, poles lie in the right half plane the system response will always contain terms like e^{at} or $e^{at} \sin(\beta t + \phi)$, where $a, \alpha > 0$. Unless special efforts are made to counteract these terms, the output will increase without limit. In particular, the output will increase without limit even if the input is removed: the system is *unstable*.

In practice, the output could only increase as much as physically possible, but with whatever dire prospects this might imply.

We now have a most important stability criterion, expressed in terms of the s-plane: for a system to be stable *all* its poles must lie in the left half plane. Even just one right half-plane pole means instability, since the growing exponential term associated with it will eventually dominate the system response. (Poles on the $j\omega$ axis itself correspond to systems on the stability borderline: the corresponding time waveforms neither die away nor increase indefinitely with time.)

Earlier, the physical interpretation of the Laplace variable s was limited to the relationship between a transfer function $G(s)$ and the frequency response function $G(j\omega)$. The preceding discussion, however, leads to a more general interpretation of the real and imaginary parts of $s = \sigma + j\omega$. Each (generally complex) value of s, as illustrated by Fig. 5.5, corresponds to a time domain signal of the form $e^{\sigma t} \sin(\omega t + \phi)$. The real part σ determines the rate of decay or increase of the signal, while the imaginary part $j\omega$ determines the frequency of the associated oscillation.

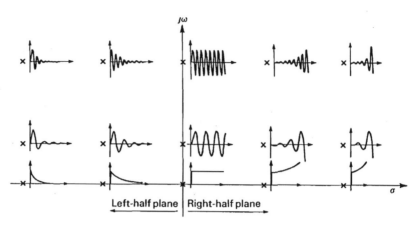

Fig. 5.5 Time waveforms corresponding to s-plane pole locations (only the upper half of the s-plane is shown). (Reproduced with the kind permission of Van Nostrand Reinhold.)

In the control context this interpretation is useful in modelling components in system response; it is also of wider relevance in signal processing applications.

Stability assessment

See Meade, M.L. and Dillon, C.R., *Signals and Systems*, Van Nostrand Reinhold, 1986 (2nd edition 1991).

The stability criterion developed above states that all system poles must lie in the left half plane for a stable system, or equivalently, all the roots of the denominator $D(s)$, of the system transfer function must have negative real parts. Because of the significance of this, the equation $D(s) = 0$ is known as the *characteristic equation*.

Often, however, we do not know exactly where the poles of a system model actually lie. Consider the general closed-loop system of Fig. 5.6, for example. The closed-loop transfer function takes the familiar form

$$\frac{C(s)}{R(s)} = \frac{KG(s)}{1 + KG(s)H(s)}$$

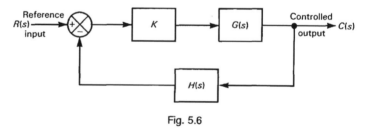

Fig. 5.6

94

with the poles located where $1 + KG(s)H(s) = 0$.

Unfortunately, even if we know where the poles and zeros of $G(s)$ and $H(s)$ lie, those of the overall closed-loop system can only be located by solving the characteristic equation $1 + KG(s)H(s) = 0$, and this may well be of a higher order, and time-consuming to solve. The stability of the system could be assessed, however, if we knew whether any of the roots of the equation have positive real parts – even without knowing the roots themselves. Can this be done?

The problem greatly exercised 19th-century mathematicians, although it was not expressed in terms of the s-plane. The criterion which emerged is known as the Routh–Hurwitz criterion after the two mathematicians who independently developed equivalent versions of it. Rather than go into details here, it is described in Appendix 2. Like many classical control engineering techniques it has declined in importance somewhat since the wide availability of control system design software, which usually includes polynomial root-finding routines. The Routh–Hurwitz criterion is still a useful tool, however, with which any competent control engineer should be familiar.

Steady-state and transient response

It is convenient to view the complete response of a linear system to any applied input as consisting of two distinct parts. Firstly, as already seen, there will be components whose general forms are determined by the poles of the system. In a stable system with poles in the left half plane these components will eventually die away.

This part of the response is therefore often known as the *transient response*. After sufficient time the transient response will be negligibly small and the system is said to be in a *steady state*. The remainder of the complete response is therefore known as the *steady-state response*, and its general form is determined only by the form of the system input. In Worked Example 5.4(a) the steady-state part of the response was the constant term $\frac{1}{2}$, while the transient response was $-\frac{1}{2}e^{-2t}$ (both for $t > 0$). Note again, however, that the steady-state response need not be unchanging in time (as we saw in Chapter 3 when discussing steady-state sinusoidal response). This is illustrated in the following example.

The general forms of the response terms derived in the previous section hold only for *distinct* poles – that is, single roots of the denominator polynomial $D(s)$. In the case of repeated roots, the general forms of the transient response terms are slightly different, but are still characterized by overall exponential decay for poles to the left of the $j\omega$ axis, and exponential increase for those to the right.

Evaluate the total response of the first-order system $G(s) = 1/(1 + s\tau)$ to a unit ramp input $x(t) = t$ applied at time $t = 0$. Identify the steady-state and transient response terms and sketch the complete response.

Worked Example 5.5

Solution   Transforming $x(t)$ we have $X(s) = 1/s^2$.
Hence the output $y(t)$ is defined by $Y(s) = G(s) \times X(s)$ – that is,

$$Y(s) = \frac{1}{s^2(1 + s\tau)}$$

An appropriate partial fraction expansion is

$$\frac{1}{s^2(1 + s\tau)} = \frac{As + B}{s^2} + \frac{C}{(1 + s\tau)}$$

Cross-multiplying gives

$$1 = (As + B)(1 + s\tau) + Cs^2$$
$$= s^2(C + A\tau) + s(A + B\tau) + B$$

Substituting $s = 0$ in the first expression gives $B = 1$.

Substituting $s = -(1/\tau) \Rightarrow C = \tau^2$.

Equating coefficients of s^2 gives $0 = \tau A + C$. Hence

$$0 = \tau A + \tau^2$$
$$A = -\tau$$

The complete response is therefore given by the transform

$$Y(s) = \frac{1}{s^2} - \frac{\tau}{s} + \frac{\tau^2}{1 + s\tau}$$

$$= \frac{1}{s^2} - \frac{\tau}{s} + \tau\frac{1}{s + 1/\tau}$$

Looking up each term in Table 5.1 leads to an expression for $y(t)$:

$$y(t) = \underbrace{t - \tau}_{\substack{\text{steady-} \\ \text{state} \\ \text{response}}} + \underbrace{\tau e^{-t/\tau}}_{\substack{\text{transient response,} \\ \text{corresponding to} \\ \text{system pole at } s = -1/\tau}} \quad (t > 0)$$

The complete response is sketched in Fig. 5.7. Note that the steady-state response is itself a ramp function, like the input, but that for this particular first-order system with unity steady-state gain it has a slope of 1, and hence lags behind the input in the steady state by a constant value τ.

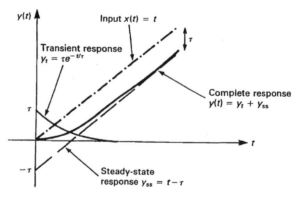

Fig. 5.7 First-order unit ramp response.

Standard models and the s-plane

A pole–zero plot of a system, as we have seen, summarizes all the most important features of linear system behaviour (apart from an undefined overall gain term) in a particularly simple form. By visualizing the three-dimensional $|G(s)|$ plot and the cut along the $j\omega$ axis, a good idea of the amplitude frequency response may often be obtained. By noting the positions of the poles, the general forms of the transient response components may be identified. And we can tell at a glance if a system is stable or unstable.

At this point it is useful therefore to re-interpret the standard models of Chapter 3 in terms of the s-plane. Some aspects of what follows have been introduced implicitly earlier in this chapter, but it is worth making a more comprehensive list here.

(1) First-order lag

$$G(s) = \frac{k}{1 + s\tau}$$

Rewriting

$$G(s) = \frac{k/\tau}{s + 1/\tau}$$

we see that the transfer function corresponds to a single pole at $s = -(1/\tau)$, as was illustrated earlier in Fig. 5.4. Note that a large value of τ corresponds to a pole near the $j\omega$ axis and a transient response taking a relatively long time to die away. A small time constant means a pole further to the left and a faster transient response.

Note also that a transfer function $G(s) = K/(s + a)$ corresponds to a time constant $\tau = 1/a$ *and a steady-state gain K/a, not K,* as can be seen by rewriting $G(s)$ in standard form

$$G(s) = \frac{K/a}{1 + s/a} \equiv \frac{k}{1 + s\tau}$$

(2) Integrator

$$G(s) = \frac{K}{s}$$

This corresponds to a single pole at the origin of the s-plane – that is, at $s = 0$.

Hence the pole at the origin of the pole–zero plot for the P + I controller in Fig. 5.2(a). Note, however, the zero which also features in the controller plot.

(3) Second-order lag

$$G(s) = \frac{k\omega_n^2}{s^2 + 2\zeta\omega_n s + \omega_n^2}$$

The poles are located where

$$s^2 + 2\zeta\omega_n s + \omega_n^2 = 0$$

That is, where

$$s = -\zeta\omega_n \pm \sqrt{\zeta^2\omega_n^2 - \omega_n^2} = -\zeta\omega_n \pm \omega_n\sqrt{\zeta^2 - 1}$$

97

There are three possibilities, therefore:

(i) overdamped $\xi > 1$

$$s = -\zeta\omega_n \pm \omega_n \sqrt{\zeta^2 - 1}$$

That is, both poles are real, and the pole–zero plot is shown in Fig. 5.8(a).

(ii) Critically damped $\zeta = 1$

In this case the denominator of the transfer function has a double root, and the system possesses a double pole at $s = -\omega_n$ (Fig. 5.8(b)).

(iii) Underdamped $0 < \zeta < 1$

Here the poles are located at the conjugate positions

$$s = -\zeta\omega_n \pm j\omega_n \sqrt{1 - \zeta^2}$$

as shown in Fig. 5.8(c).

Figure 5.8 also shows the typical step responses associated with each case. For the overdamped system, $\zeta > 1$, both poles are real and the transient part of the response is non-oscillatory. As ζ is decreased for a given ω_n, the rise time of the system decreases. Fastest rise time without overshoot occurs for $\zeta = 1$ – that is, when there is a double pole at $s = -\omega_n$. For underdamped systems, $0 < \zeta < 1$, the poles become a complex conjugate pair, implying a transient response com-

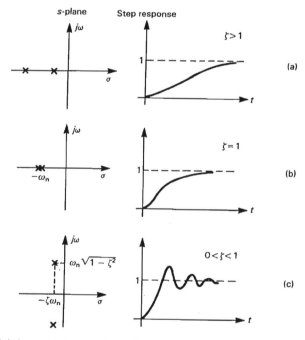

Fig. 5.8 Relationship between pole positions and step response for a second-order lag.

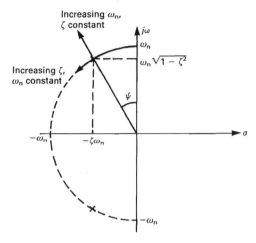

Fig. 5.9 Damping line and damping angle.

ponent of the form $Re^{-\zeta\omega_n t}\sin(\omega_n \sqrt{1-\zeta^2}t + \phi)$ – that is, an exponential envelope of the form $Re^{-\zeta\omega_n t}$, determined by the real part of the pole positions and a frequency of oscillation $\omega_{osc} = \omega_n \sqrt{1-\zeta^2}$ determined by the imaginary part. Such components give rise to overshoot and an increasingly oscillatory step response as ζ is decreased.

Refer to Fig. 3.21 for a reminder of step responses for different values of ζ.

It is useful to introduce a further concept for underdamped second-order systems, that of the *damping line* illustrated in Fig. 5.9. This is the line drawn through the pole position and the origin of the s-plane; similarly the angle between the damping line and the $j\omega$-axis is known as the *damping angle* ψ. From Fig. 5.9 it may easily be seen that

$$\sin\psi = \frac{\zeta\omega_n}{\omega_n} = \zeta$$

Hence

$$\zeta = \arcsin\psi$$

That is, the damping angle depends only on the damping ratio of the system: all poles lying on a given damping line will have the same damping ratio $\zeta = \arcsin\psi$. The natural undamped angular frequency ω_n will vary from system to system, however. Specifically, ω_n is equal to the radial distance of the poles from the origin, as shown in Fig. 5.9. The rate of decay of the transient response, determined by the real part $-\zeta\omega_n$ of the pole position, will depend on both ζ and ω_n.

A knowledge of ζ alone allows us to estimate the form but not the time scale of transient behaviour. For the time scale we need ω_n as well – hence the time-normalization of second-order step response with respect to ω_n.

(4) Time delay

As already noted, the transfer function corresponding to a time delay is not in the form of a ratio of polynomials. In fact, it takes the form

$$G(s) = e^{-sT}$$

where T is the pure time delay.

This expression may be derived formally from the properties of the Laplace transform: for our purposes, however, we shall simply note its similarity to the corresponding frequency response function derived in Chapter 3:

$$G(j\omega) = e^{-j\omega T}$$

A finite pole–zero approximation to the transfer function $G(s) = e^{-sT}$ may be obtained in a number of ways, by first expanding the exponential function as a series and then seeking simple rational polynomial approximations; those interested in such techniques are referred to the texts listed as 'Further Reading'.

There is no simple pole–zero plot corresponding exactly to the transfer function $G(s) = e^{-sT}$.

Hence it is often easier to treat systems with time delay in the frequency domain, as was the case with the design example of Chapter 4; a pure time delay *can* be represented exactly on a Bode plot.

Higher-order systems and dominance

Consider the system with the transfer function

$$G_1(s) = \frac{1}{(s + 1)(s + 5)(s + 10)}$$

In general, the transient response of this system will take the form

$$K_1 e^{-t} + K_2 e^{-5t} + K_3 e^{-10t}$$

The general shape of these exponential functions is shown in Fig. 5.10, which demonstrates how quickly e^{-5t} and e^{-10t} die away compared with e^{-t}. Even if the constant coefficients K_3 or K_2 associated with the terms e^{-10t} or e^{-5t} are appreciably larger than K_1, the nature of the exponential function ensures that after a relatively short period of time the term in e^{-t} completely dominates the transient response. In such cases the corresponding pole (in this example the pole at $s = -1$) is termed the *dominant pole*.

A similar argument holds for complex conjugate pairs of poles. The system G_2 with transfer function

$$G_2(s) = \frac{1}{(s + 5)(s^2 + 2s + 5)}$$

has poles at $s = -5; -1 \pm 2j$. The transient response associated with the real pole is of the form Ke^{-5t}, while that associated with the complex pair is of the form

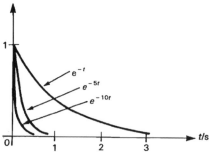

Fig. 5.10

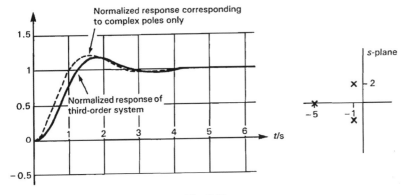

Fig. 5.11

$Re^{-t}\sin(2t + \phi)$. This time, the Ke^{-5t} term will die away much faster than the term $Re^{-t}\sin(2t + \phi)$, leaving the latter to dominate the response. Here, therefore, there is a dominant complex conjugate pair of poles.

The notion of a dominant pole or conjugate pair of poles is useful in design, as will be seen in the next chapter, since it often allows the behaviour of a higher-order system to be approximated by the standard model corresponding to the dominant pole or pair of poles. For example, the step response of the system G_2 illustrated in Fig. 5.11 is very similar to that corresponding to the complex poles alone (normalized to the same steady-state value). If the dominant poles are closer to the $j\omega$ axis than any other poles by more than a factor of, say, 5–10, the overall system response may often be approximated by the response of the dominant pole(s) alone. However, like so many other techniques in control engineering, there are no hard-and-fast rules about when poles are sufficiently dominant to allow such approximations to be made. All we can guarantee is that the real part of a pole position determines the rate of decay of the associated term in the transient response of a stable system. The weighting of the individual terms, and hence the *precise* form of the response, is determined by the *complete* transfer function together with the specific input, and there are cases where the simple idea of dominance needs to be qualified. I shall briefly examine two special cases here: (a) the case of dominant complex poles corresponding to a low damping ratio ζ and (b) the effect of an additional real zero on step response. The examples are constructed artificially to exaggerate the effects described, but the general principles are important for practical design.

(a) System with low damping

Consider the system modelled by the transfer function

$$G_3(s) = \frac{1}{(s + 5)(s^2 + 2s + 26)}$$

with poles located at $s = -5; -1 \pm 5j$. In other words, a system similar to G_2, but with complex poles at $-1 \pm j5$ instead of $-1 \pm j2$. Applying the previous argument, we would expect the transient response to be dominated by the oscillatory

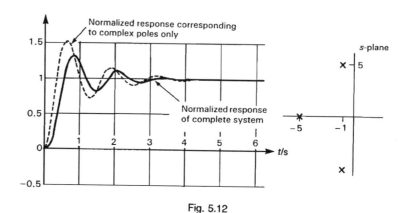

Fig. 5.12

component – as indeed it is. However, in this case we *cannot* assume that the step response of the third-order system is almost identical to that of the second-order system represented by the complex poles alone. Figure 5.12 illustrates simulated step responses, normalized to identical steady-state values. The complex poles alone correspond to a damping ratio ζ of around 0.2 and an overshoot of over 50%. The third-order system including the real pole at $s = -5$, on the other hand, exhibits an overshoot of only about 35%, as well as a rather longer rise time. The effect of the real pole has been to slow down the response and to decrease the overshoot, despite a difference of a factor of 5 in the real parts of the pole positions. The same factor of 5 for the system G_2, with more heavily damped complex poles, resulted in a much smaller difference in the two step responses of Fig. 5.11.

The effect of the additional real pole can be appreciated in the following way. Because of the nature of transfer function models, the behaviour of system G_3 is identical to that of either of the cascaded systems in Fig. 5.13. In Fig. 5.13(a) the real pole can be interpreted as an additional first-order lag which smooths out the initial rise of the input step: overall response is thereby slowed down and the 'ringing' of the underdamped second-order system is reduced. Alternatively, we

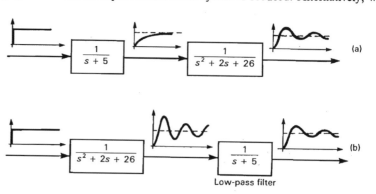

Fig. 5.13 Effect of an additional real pole on an underdamped second-order system.

102

can adopt a frequency domain approach, and view the extra first-order lag as a low-pass filter acting on the output of the second-order term, as in Fig. 5.13(b). Then the filtering effect of the real pole reduces the bandwidth of the overall system. If the cut-off frequency $1/\tau$ associated with the real pole is of a similar magnitude to the frequency of the ringing, the filtering will significantly affect overall behaviour. In this particular case $1/\tau = \omega_{\text{osc}} = 5\,\text{rad s}^{-1}$ and the effect is substantial.

The bandwidth of a closed-loop system is often used as a general performance measure in much the same way as natural frequency. Since virtually all control systems are low-pass in nature, bandwidth is usually defined as the frequency above which the amplitude ratio is less than a given fraction of the low-frequency value. The fraction chosen is most commonly 3 dB below the low frequency gain.

(b) Effect of a real zero

Consider the system with the transfer function

$$G_4(s) = \frac{(1 + 0.5s)}{(s^2 + 2s + 5)}$$

with a zero at $s = -2$ and complex poles (like G_2) at $s = -1 \pm j2$.

A control system with a similar transfer function will be considered in Chapter 6.

The transform of the step response of this system is given by

$$\frac{G_4(s)}{s} = \frac{1 + 0.5s}{s(s^2 + 2s + 5)} = \frac{1}{s(s^2 + 2s + 5)} + \frac{0.5}{(s^2 + 2s + 5)}$$

This consists of two terms. The first corresponds to the step response of a second-order lag with complex poles alone, at $s = -1 \pm j2$. The second term corresponds to an exponentially decaying sinusoid of the form $Ke^{-t}\sin 2t$ (from Table 5.1). The effect of adding in this second term to the step response is to decrease the rise time and increase the overshoot of the system with the zero, as shown in Fig. 5.14. (Step responses have again been normalized to a steady-state value of 1.) Although the complex poles alone would indicate a step response with an overshoot of less than 25%, the system G_4 with the added zero overshoots its steady-state value by more like 40%.

The effect of the zero can also be appreciated in frequency domain terms. In this case, we could imagine the term $(1 + 0.5s)$ to be detached from the rest of the system G_4. The corresponding frequency response function is $(1 + 0.5j\omega)$, and has the effect of boosting high frequencies and increasing the system bandwidth.

You can check this by sketching the Bode plot of the term $(1 + 0.5j\omega)$, which is similar to what was called the 'lead' term in the P + I frequency response described in Chapter 4.

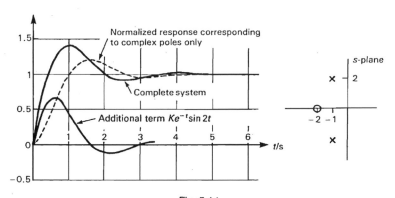

Fig. 5.14

An increase in bandwidth implies a faster time domain response, owing to the increased effect of the high-frequency components. This is reflected in the faster rise time of the system with the extra zero.

In all three cases – the basic system G_2, the more lightly damped system G_3 and the system G_4 with the added zero – the dominant transient response term dies away at the same rate, determined by the same real part of the dominant complex poles. The relative weightings of the transient response terms, however, differ sufficiently to give the considerably different complete step responses. To sum up:

(a) The transient response term corresponding to the pole (or complex conjugate pair of poles) nearest to the $j\omega$-axis eventually dominates the transient response of the system.
(b) In many cases, the behaviour of a complex, higher-order system may be approximated by the behaviour of a system corresponding to the dominant pole(s) alone.
(c) If a system is dominated by 'lightly damped' complex poles, then the influence of apparently non-dominant poles may be significant.
(d) Zeros located relatively close to dominant poles can greatly influence time domain performance. (Zeros *closer* to the $j\omega$-axis than the dominant pole(s) can have an overwhelming effect.)

For a more detailed discussion of dominance see Kuo, B.C., *Automatic Control Systems*, 5th ed., Prentice-Hall, 1987.

Worked Example 5.6 A system has a transfer function

$$G(s) = \frac{1320}{(s + 8)(s^2 + 10s + 61)(s^2 + 2s + 2.7)}$$

(a) Sketch the pole–zero plot of the system.
(b) Identify the dominant pole(s) and the associated system parameter(s) (τ, ζ, ω_n as appropriate).
(c) Sketch the step response (normalized to a steady-state gain of 1) of a system consisting of the dominant pole(s) alone. If you have appropriate computer software available, simulate the step response of the complete, fifth-order system.

Solution
(a) The poles are located at

$$s = -8$$
$$s = -5 \pm j6$$
$$s = -1 \pm j1.3$$

and are sketched in Fig. 5.15(a).
(b) The dominant poles are those at $s = -1 \pm j1.3$. The corresponding ζ and ω_n may be estimated by considering the damping line and damping angle, or immediately from the factor $(s^2 + 2s + 2.7)$ in the transfer function. Setting

$$s^2 + 2s + 2.7 \equiv s^2 + 2\zeta\omega_n s + \omega_n^2$$

gives $\omega_n^2 = 2.7$; $\omega_n = 1.64 \, \text{rad s}^{-1}$; $\zeta\omega_n = 1$; $\zeta = 0.61$.
(c) The second-order step response corresponding to $\zeta = 0.61$, $\omega_n = 1.64$ and a steady-state value of 1 is shown in Fig. 5.15(b) and compared with a simulated step response of the complete system. In this case the dominant poles are

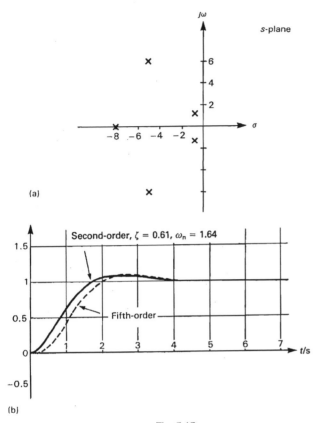

(a)

Second-order, $\zeta = 0.61$, $\omega_n = 1.64$

Fifth-order

(b)

Fig. 5.15

closer to the $j\omega$ axis by a factor of 5 than the second conjugate pair, and there is a similar factor in their relative distances from the origin. There are no zeros in the vicinity of the dominant poles, so it is reasonable to assume that the complete system response is approximated fairly well by the step response corresponding to the dominant poles – as is demonstrated by simulation. The main effect of the additional poles has been to slow down the response somewhat, but overshoot is almost unaffected.

Matching frequency response curves

It is sometimes possible to approximate higher-order systems by a 'best-fit' match of the frequency response amplitude plot to a standard second-order model, in a way similar to that carried out for the design exercise of Chapter 4. In such cases the 'best-fit' values of ω_n and ζ will not necessarily correspond to those suggested

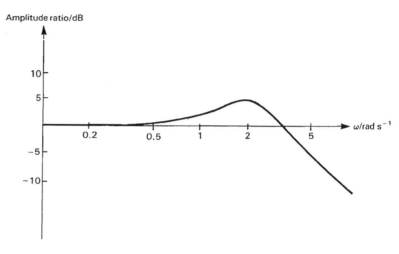

Amplitude ratio/dB

Fig. 5.16

by any particular individual pair of poles of the higher-order system. Matching frequency responses in this way does not select a particular pair of poles as dominant, it merely indicates values of ω_n and ζ with which to approximate overall system behaviour. The degree to which the approximation holds in practice depends on the degree to which the amplitude and phase plots of the higher-order system deviate from those of the second-order approximation over the frequency range of interest.

For example, Fig. 5.16 shows the amplitude frequency response of the system G_4 with the complex poles and the real zero. The plot has again been normalized to unity low-frequency gain. Although the amplitude plot will not fit closely any second-order model (particularly at high frequencies, where the effect of the zero reduces the slope of the roll-off), it seems reasonable to approximate the general behaviour by that of a second-order system, and take the height of the resonance peak as a measure of the effective damping. Figure 5.16 shows a peak of about 5 dB in the amplitude ratio. Using the standard curves of Chapter 3, this corresponds to a damping ratio of 0.3 and a step response overshoot of just over 35% – a much closer match to the simulated step response of Fig. 5.14 than the step response of a system with the complex poles alone.

Summary

Differential equation models may be used to derive a transfer function $G(s)$ in the Laplace variable, s. System models in terms of s give a concise representation of (i) frequency response $G(j\omega)$ (by substituting $s \rightarrow j\omega$); (ii) transient response (general form determined by pole positions); and (iii) stability (all poles are in the left half of the s-plane for stable systems).

A system's frequency response shows the influence of all the system poles and zeros (think of the $|G(s)|$ surface and the cut along the $j\omega$ axis). Matching frequency response in this way can therefore sometimes give a better approximation to overall behaviour than considering only the dominant poles.

106

The concept of dominant poles allows useful approximations of system behaviour to be made. The transient response of a system is dominated by the terms associated with the poles nearest the $j\omega$-axis, whose real parts determine how long the dominant transient response terms take to die away. The complete time domain response of a system may often be approximated by the response associated with the dominant pole(s) only. Care must be exercised, however, in the case of lightly damped systems, and systems possessing a zero in the vicinity of the dominant pole(s).

Problems

5.1 Use Table 5.1 to find the Laplace transforms of the following signals:
   (a) $e^{-2t} + e^{-3t}$
   (b) $2 - e^{-t}$
   (c) $\sin 5t$
   (d) $e^{-4t} \cos 2t$.

5.2 Use Table 5.1 to find the signals corresponding to the following transforms:
   (a) $6/(s + 2)$
   (b) $5/(1 + 10s)$
   (c) $1/(s^2 + 4s + 4)$
   (d) $1/(s^2 + 4s + 5)$.

5.3 (a) Use partial fractions to find the signal $f(t)$ corresponding to the transform

$$F(s) = \frac{3s + 12}{s(s + 2)(s + 3)}$$

   (b) A signal $f(t)$ has the transform

$$F(s) = \frac{s + 6}{s^2 + 2s + 26}$$

   Express $f(t)$ in the form $R \exp(-at) \sin(\omega t + \phi)$.

5.4 Plot the pole–zero diagrams corresponding to the following transfer functions. In each case state whether the corresponding system is stable. For stable systems, also state the general form of the associated transient response.

   (a) $\dfrac{10}{(s + 2)(s + 0.3)}$

   (b) $\dfrac{s}{s - 1}$

   (c) $\dfrac{1 + s}{(1 + 5s)(1 + 10s)}$

   (d) $\dfrac{1}{s^2 + 1.5s + 10}$

   (e) $\dfrac{1}{s^2 + 4}$

(f) $\dfrac{5}{s(s + 6)}$

5.5 What are the steady-state gains of
(a) the systems with transfer functions

$$G_1(s) = \frac{5}{s + 2}$$

and

$$G_2(s) = \frac{50}{(s + 2)(s^2 + 2s + 10)}$$

(b) the fifth-order system of Example 5.6.

5.6 Sketch the amplitude ratio $|G_3(j\omega)|$ corresponding to the transfer function $G_3(s)$ given on page 101. By matching this amplitude–frequency response curve to a standard second-order model, suggest appropriate values of ζ and ω_n with which to approximate the third-order system. Compare the step response suggested by your second-order model with Fig. 5.12.

5.7 Consider again the system of Fig. 4.30(a), where the proportional gain K of the controller is 5. Assume that $\theta_i = 0$, but that the disturbance d varies with time. Calculate the *disturbance transfer function* $\theta_o(s)/D(s)$ under these conditions for (a) $T_i = 4$s and (b) $T_i = 10$s. Hence derive analytical forms for the responses to a unit step change in disturbance illustrated in Fig. 4.30(c).

The root-locus technique   6

☐ To explain the fundamentals of the root-locus technique.
☐ To indicate how the root-locus method may be used to select controller gain.
☐ To present guidelines for sketching the root loci of simple systems.
☐ To describe, with the aid of simple root loci, the effect of velocity feedback and proportional plus derivative control.

Objectives

These objectives are deliberately limited in scope. Expertise in the material presented here comes only with much practice: this chapter should be seen simply as an introduction to the root-locus technique, laying a foundation for subsequent more detailed study if desired.

The previous chapter described how the roots of a system's characteristic equation – in other words, the system poles – determine the general form of the transient response terms. This close relationship between pole position and time domain behaviour provides the basis of a major design technique for single-input, single-output feedback control systems. The essence of the technique is a graphical method of indicating on the s-plane all the possible closed-loop pole locations as some parameter (most usually the controller gain) is varied. The designer can then select a value of this parameter so as to position the closed-loop poles at appropriate positions in the plane. Most importantly of all, perhaps, this locus of all possible *closed-loop* pole positions can be derived from a knowledge of the *open-loop* transfer function alone.

The basic idea has already been introduced in Chapter 4, where the effect on closed-loop behaviour of changing controller gain was investigated for simple first- and second-order processes. The frequency response results obtained there can easily be reinterpreted in terms of so-called *root loci* on the s-plane.

First- and second-order root loci

Consider again the motor speed control system of Chapter 4, illustrated in Fig. 6.1. The motor is represented by the first-order transfer function

$$G(s) = 0.5/(s + 1)$$

– that is, there is a single open-loop pole at $s = -1$. The closed-loop transfer function is given as usual by

$$\frac{KG(s)}{1 + KG(s)} = \frac{0.5K/(s + 1)}{1 + 0.5K/(s + 1)}$$
$$= \frac{0.5K}{s + (1 + 0.5K)}$$

The closed-loop system is therefore also first order, with the pole lying at $s = -(1 + 0.5K)$.

This expression indicates very clearly how the system closed-loop pole position depends on the controller gain, K. When K is very small, the term $(1 + 0.5K)$ is

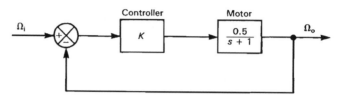

Fig. 6.1 The motor speed control system.

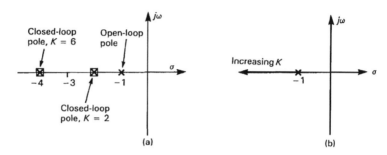

Fig. 6.2 The speed control system root locus.

very nearly equal to 1, and the closed loop pole lies near to the open-loop pole position at $s = -1$. As K is increased, the closed-loop pole is located increasingly to the left on the σ-axis, as shown in Fig. 6.2(a). For $K = 2$, for example, the pole lies at $s = -2$; for $K = 6$, at $s = -4$.

The locus of all possible *closed-loop* pole positions is therefore a line running left along the σ-axis starting at the *open-loop* pole position, as illustrated by Fig. 6.2(b). *Root-locus* plots such as this conventionally show the open-loop poles and zeros, as well as the lines representing all possible positions of closed-loop poles.

Arrows are used to indicate on the locus lines how the closed-loop pole positions change as the relevant parameter is increased. Exactly the same general features can be deduced from the s-plane root-locus plot of Fig. 6.2(b) as were derived in Chapter 4 from the variation of closed-loop frequency response – namely, that the closed-loop system is also first order, with a time constant which decreases progressively as the gain is increased.

Remember that the time constant of a first-order system has a magnitude equal to the inverse of the pole position.

The root locus of the position control system of Fig. 6.3 can be derived in a similar way. Following exactly the same procedure as with the frequency response function of Chapter 4, but with s replacing $j\omega$, the closed-loop transfer function is:

$$\frac{\theta_o(s)}{\theta_i(s)} = \frac{0.5K}{s^2 + s + 0.5K}$$

The closed-loop poles therefore lie where

$$s^2 + s + 0.5K = 0$$

that is, where

$$s = \frac{-1 \pm \sqrt{1 - 2K}}{2}$$

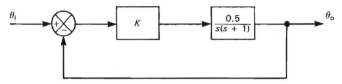

Fig. 6.3 The position control system.

When K is very small, the closed-loop poles are again very near the open-loop poles at $s = 0$ and -1. As K is increased slightly, but kept below 0.5, the closed-loop poles move closer together along the real axis between 0 and 1.

For $K > 0.5$ the closed-loop poles are complex, located at

$$s = \frac{-1 \pm j\sqrt{2K - 1}}{2}$$

Under these circumstances, then, the real part is always $\sigma = -0.5$, while the imaginary part increases as K is increased: for $K > 0.5$, therefore, the closed-loop poles lie on a line parallel to the $j\omega$-axis defined by $\sigma = -0.5$. Figure 6.4 shows the complete root locus for this system, together with the closed-loop pole positions for various values of gain.

Note how the root-locus plot alone summarizes the important features of how closed-loop behaviour changes as the gain is increased. When the gain is very low, the closed-loop step response will be dominated by a pole near the origin. Increasing the gain a little results in a faster step response, since the dominant pole will now be located further to the left. (At the same time, the other closed-loop pole on the real axis moves to the right.) A further increase in gain leads eventually to a critically damped system with a double pole at $s = -0.5$. Higher

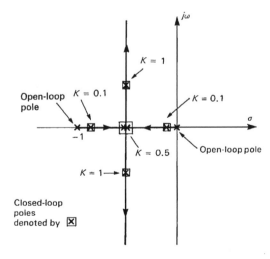

Fig. 6.4 The position control system root locus.

gain still results in complex poles, and a progressively more underdamped system. The response will now include oscillatory components whose frequency will increase with increasing K, as the complex closed-loop poles move progressively further from the σ-axis. This simple example demonstrates the power of the root locus in representing a considerable amount of information about the system in extremely compact form.

Worked Example 6.1

Sketch the root loci of the unity feedback, proportional control systems where the plant is modelled as

(a) an integrator $G(s) = 1/s$; and
(b) a double integrator $G(s) = 1/s^2$.

Solution

(a) The closed-loop transfer function takes the form

$$\frac{K/s}{1 + K/s} = \frac{K}{s + K}$$

As K is increased, therefore, the closed-loop pole lies further to the left along the negative real axis. The root locus is sketched in Fig. 6.5(a).

(b) In this case the closed-loop transfer function is

$$\frac{K/s^2}{1 + K/s^2} = \frac{K}{s^2 + K}$$

The denominator of this expression is zero when $s^2 = -K$. For positive controller gain, this means that the poles must lie on the $j\omega$-axis, at $s = \pm j\sqrt{K}$. They move further from the origin as the gain is increased. The root locus is therefore as shown in Fig. 6.5(b).

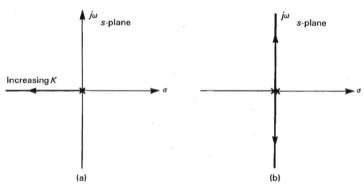

(a) (b)

Fig. 6.5

An alternative approach

The method just described of calculating root loci suffers from one major problem: to sketch the loci, we had effectively to solve the characteristic equation. There is, however, an alternative method, requiring only a knowledge of the open-loop pole–zero configuration.

In many practical systems, the complete loop is made up of a number of cascaded individual components. Figure 6.6(a) shows such a system for controlling the position of a mechanical structure (such as the telescope antenna of Chapter 3, for example), with transfer function models given for each component. Calculating the location of the closed-loop poles of this system for any given value of gain K would involve solving a fourth-order characteristic equation, which can be very time-consuming, even with computer assistance.

The open-loop transfer function, however, is made up of the transfer functions of the individual components multiplied together, so the corresponding open-loop pole–zero diagram is obtained simply by plotting the poles and zeros of the individual components on a single diagram (Fig. 6.6(b)). Open-loop pole–zero plots can often be obtained in this way with very little computation, even for

The equation will be of even higher order if transducer dynamics have to be taken into account as an additional transfer function in the feedback path.

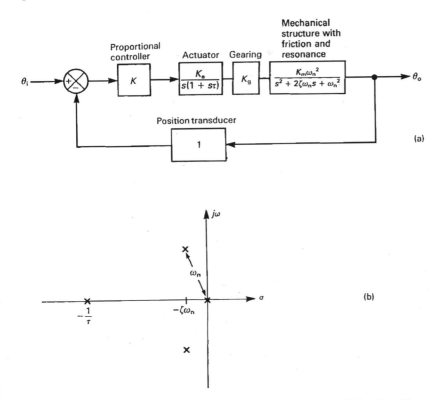

Fig. 6.6 A system for position control of a mechanical structure with inertia, stiffness and damping.

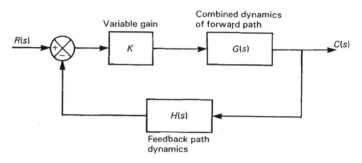

Fig. 6.7

higher-order systems. A technique for deducing features of the root locus based only on such open-loop pole–zero configurations is therefore very attractive.

In the days before widely available control system design software, control engineers needed to develop a high degree of expertise in constructing accurate root loci. Nowadays, it may be argued, the value of root loci lies rather in the convenient way they allow the engineer to assess quickly the feasibility of a proposed control scheme, or to appreciate at a glance the general form of dynamic behaviour which such a scheme might imply. Computer-based tools are almost indispensable for the detailed design of complex systems using root-locus methods. Such tools can only be effective, however, when the user has a good fundamental understanding of the general nature of the system under considera-tion. Such an understanding can be greatly enhanced by an elementary knowledge of the root-locus technique.

Consider the closed-loop transfer function of the general system of Fig. 6.7:

$$\frac{C(s)}{R(s)} = \frac{KG(s)}{1 + KG(s)H(s)}$$

The closed-loop poles lie where

$$1 + KG(s)H(s) = 0$$

That is, where $G(s)H(s) = -1/K$.

For this latter condition to hold with a positive value of K, the angle, or *phase* $\angle G(s)H(s)$ of the open-loop transfer function, must be $\pm 180°$ or an odd multiple thereof. Only then can the complex expression represented by $G(s)H(s)$ take on a real, negative value. Since we are interested in all possible positive values of K, we need to investigate all possible values of s for which the phase of the open-loop transfer function satisfies this condition. When a transfer function is expressed as the product of individual pole and zero terms it is possible to do this by summing the phase contributions of all individual open-loop poles and zeros:

If $\quad G(s)H(s) = \dfrac{K(s - z_1)(s - z_2) \dots (s - z_m)}{(s - p_1)(s - p_2) \dots (s - p_n)}$

then $\quad \angle G(s)H(s) = \sum_{i=1}^{m} \angle(s - z_i) - \sum_{j=1}^{n} \angle(s - p_j)$

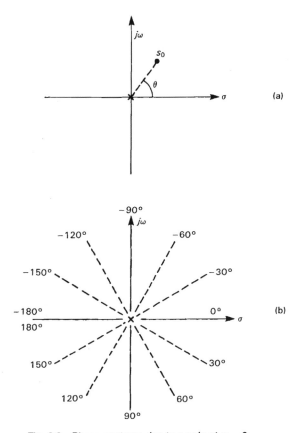

Fig. 6.8 Phase contours due to a pole at $s = 0$.

The easiest way to appreciate the consequences of this for the shape of root loci is to indicate $\angle G(s) H(s)$ directly on the s-plane as contours of equal phase. Then the root locus will follow the contours representing odd multiples of $\pm 180°$. Consider first the simplest possible case – an open-loop transfer function consisting of one pole only, located at the origin.

Figure 6.8(a) shows a general point $s = s_0$ on the s-plane, with a single open-loop pole at the origin. For $s = s_0$ the phase of the open-loop transfer function, $1/s$ is equal to $\angle 1/s_0 = -\angle s_0$, by the normal rules of complex numbers. But $\angle s_0$ is simply the angle θ shown in the figure. So we can mark the phase contribution of the pole throughout the s-plane as a set of *phase contours* radiating from the pole, as shown in Fig. 6.8(b). From this it is clear that the only region of the plane where the $\pm 180°$ phase condition is satisfied is the contour running left from the pole along the σ-axis. So the closed-loop pole must lie somewhere along this line, irrespective of the gain. This is exactly the result obtained earlier in Worked Example 6.1 from an analytic examination of the transfer function.

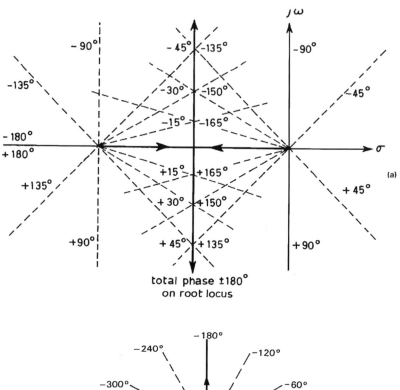

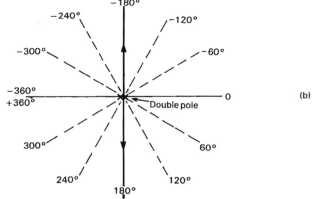

Fig. 6.9 Phase contours and root locus for $G(s)\,H(s) =$ (a) $1/s(s + 1)$ and (b) $1/s^2$.

A similar argument applies to poles lying anywhere in the *s*-plane. In each case, the phase contributions of individual poles are represented by radial contours as before. When there is more than one pole, the total phase of $G(s)\,H(s)$ for any particular value of *s* is obtained by superimposing the contours due to each open-loop pole at the corresponding point in the *s*-plane and adding the individual phase values.

This is similar to the convenient additive property of phase on the Bode plots of Chapter 4.

Viewed in this light, the other root loci considered so far can be derived

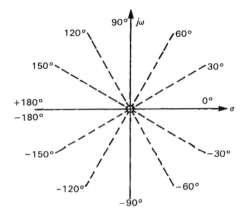

Fig. 6.10 Phase contours due to a zero at $s = 0$.

immediately from the open-loop pole positions. Figure 6.9 shows the phase contours corresponding to Figs 6.4 and 6.5(b). In each case, the root locus follows the $\pm 180°$ contour on the s-plane: the closed-loop poles *must* lie somewhere along such a line.

The phase contour method can also be applied to zeros. The contours again radiate from the zero position, but this time have opposite sign, as shown in Fig. 6.10 for a zero at the origin. A zero contributes phase *lead* where a pole contributes phase *lag*, and vice versa.

Worked Example 6.2

Figure 6.11 shows the open-loop pole–zero configuration of a system. By considering phase contours show those sections of the real axis which form part of the root locus.

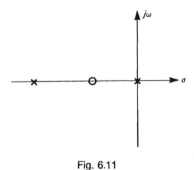

Fig. 6.11

Solution The phase contours of the two poles and the zero are shown in Fig. 6.12. The part of the negative real axis between the pole at the origin and the zero is clearly part of the root locus, since the pole at the origin contributes $-180°$ and the zero and the other pole both contribute $0°$. The region between the zero and

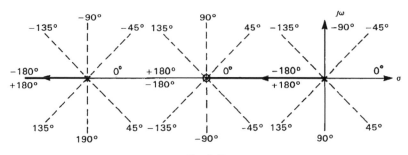

Fig. 6.12

In fact, Fig. 6.12 shows the whole of the root locus for this system, as will become clear below.

the leftmost pole is *not* part of the locus, however. The leftmost pole contributes 0°, and the contributions from the other pole and zero cancel out. Beyond the leftmost pole the first pole and zero continue to counteract each other, but total phase is now −180°, due to the new contribution of this leftmost pole.

The result of Worked Example 6.2 can be generalized to provide an extremely useful aid to sketching loci. Suppose there were a pair of complex poles in addition to the real poles and zero of Fig. 6.11. At any point on the real axis the phase contribution from either one of the complex poles would be counteracted by that of the other; the complex poles would therefore not affect the part of the root locus on the real axis. A similar argument applies to a pair of complex zeros. In fact, since only real poles and zeros contribute to the locus on the real axis, **only those sections of the real axis which lie to the left of an *odd* number of real poles plus zeros correspond to an *odd* multiple of ±180°, and hence form part of the root locus.**

In the case of higher-order systems, of course, matters become rather more complicated. Figure 6.13(a) shows a third-order system, and the corresponding root-locus diagram is given in Fig. 6.13(b). For clarity, only the total phase contours are shown, not those of the individual poles. Note that there are three branches to this third-order locus, one starting at each open-loop pole. Note also that the effect of the third pole is to push the complex branches of the root locus towards the right compared with the second-order plots of Fig. 6.9.

Remember that the poles and zeros marked on a root-locus diagram are those of the *open-loop* transfer function. The bold lines of the locus represent all possible locations for *closed-loop* poles as the gain is varied.

Worked Example 6.3

Describe in qualitative terms the general form of the transient response of the closed-loop system of Fig. 6.13 as the gain is gradually increased from a small value.

Solution For sufficiently small values of gain, there will be a closed-loop pole located near the origin, which will dominate the transient response: the system will behave very like a first-order lag. As the gain is increased, the root-locus plot shows that the two rightmost closed-loop poles will be located closer together, while the leftmost pole will lie even further to the left. The transient response terms will die away more quickly as the dominant pole moves to the left. At the

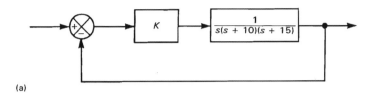

(a)

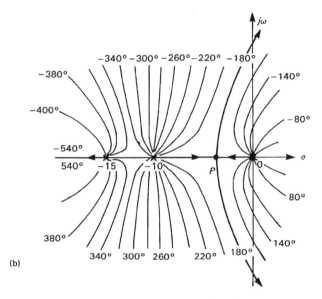

(b)

Fig. 6.13 Phase contours and root locus for a third-order system.

value of gain corresponding to the point P on the locus (known as the *breakaway point*) there will be a dominant closed-loop double pole: the system behaves like a critically damped second-order system. Further increase in gain leads to closed-loop poles on the complex branches of the root locus, and there will be corresponding oscillatory components in the transient response. Higher gain still results in these complex closed-loop poles lying farther from the σ-axis but nearer to the $j\omega$-axis; the response takes longer to reach its steady-state value and the frequency of the oscillatory components increases. With a sufficiently high value of controller gain the complex closed-loop poles have *positive* real parts (they lie on the part of the root locus which passes into the right half plane), and the closed-loop system becomes unstable.

It takes quite a lot of practice to interpret a root locus in this way. Make sure you understand all the details of this description before reading on.

Design with root locus

The above third-order example can be used to give an indication of how the root-locus technique can be used for design. For example, suppose that the specification of the third-order system of Fig. 6.13 can be satisfied if the closed-loop system

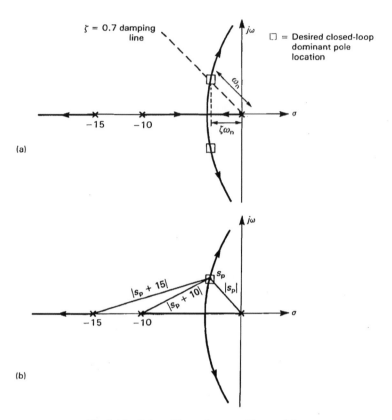

$\zeta = 0.7$ damping line

□ = Desired closed-loop dominant pole location

$j\omega$

ω_n

σ

-15 -10

$\zeta\omega_n$

(a)

$j\omega$

s_p

$|s_p + 15|$ $|s_p|$

$|s_p + 10|$

σ

-15 -10

(b)

Fig. 6.14   Gain setting using a root-locus plot.

possesses a dominant, complex conjugate pair of poles corresponding to a damping ratio of 0.7. The root-locus plot indicates that this can indeed be achieved by selecting the appropriate controller gain, since a damping ratio of 0.7 means a damping angle of 45° and the root locus does pass through an appropriate s-plane location (Fig. 6.14(a)). An accurate root-locus plot also allows other system parameters to be estimated, such as natural frequency and settling time. Returning to Fig. 6.14(a), for example, the real part of the dominant pole positions will give a measure of the settling time obtainable with a damping factor of 0.7, while their distance from the origin is equal to the natural frequency ω_n.

The required controller gain can be determined by returning to the general closed-loop pole criterion

$$G(s) H(s) = -1/K.$$

Until now we have been concerned only with the restriction this places on *phase*. However, it also places a *magnitude* constraint on all locations on the root locus, namely

$$|G(s) H(s)| = 1/K,$$

If these values turn out not to satisfy the specification, root-locus design consists in determining a suitable controller in order to force the root locus to pass through a more suitable region of the s-plane instead. This problem will be considered later.

and this expression can be used to determine the gain K required to place the dominant closed-loop poles at the desired position. The expression may be evaluated algebraically, by substituting in the value of s at the desired position, or graphically on the root-locus plot. The graphical method will be presented briefly since it is still a useful technique for making quick estimates of gain, even when computer tools are available.

In the present unity feedback example $H(s) = 1$ and

$$G(s) = 1/s(s + 10)(s + 15)$$

Hence the magnitude condition for those values of $s = s_p$ corresponding to the closed-loop pole positions is

$$|G(s_p)| = 1/K$$

That is,

$$K = |s_p| \, |s_p + 10| \, |s_p + 15|$$

Now, using the normal rules of complex number addition, each of these terms is represented by the length of a line from the test point s_p to the appropriate open-loop pole, as shown in Fig. 6.14(b). Measuring these lines (using the scale of the plot) gives

$$K = 4.8 \times 7.7 \times 12 \approx 440$$

A value of proportional controller gain has now been selected to place the dominant closed-loop poles at the desired location. If there had been any open-loop zeros, the gain-setting calculation would also have included appropriate zero vectors, such that

$$K = \frac{\text{product of lengths of vectors from poles to } s_p}{\text{product of lengths of vectors from zeros to } s_p}$$

The preceding example assumed that an accurate root-locus plot was available, although we have not yet examined how such plots may be obtained in general. In order to pursue this approach to controller design any further we therefore need to investigate the nature of higher-order root loci in more detail.

The calculation assumes that the numerator of the process transfer function is equal to 1. This was the case in Fig. 6.13(a), but is unlikely to be so in any real system. The same root locus would be obtained, however, for any constant factor in the numerator of the open-loop transfer function. When setting controller gain, the value of K calculated from the pole vectors must be reduced by an appropriate amount to take into account multiplier factors in the transfer functions of the various components.

Sketching simple root loci

The root loci of many simple systems may be sketched fairly easily if the $\pm 180°$ phase contour rule is carefully applied. In this section I shall use a simple example to illustrate a step-by-step procedure, and I shall try to draw general conclusions from it where possible. The texts listed under 'Further Reading' all give lists of more detailed, traditional 'rules' for constructing root loci. However, even these fuller guidelines do not enable the fine details of higher-order root loci to be sketched without considerable experience of commonly occurring patterns. If the root-locus approach is to be used for the design of higher-order systems, such fine details are nowadays best left to a computer package. What *is* important, however, is for the control engineer to become familiar with the general features of common patterns, and to understand how modifications to a given open-loop transfer function are likely to affect the corresponding root locus.

Before working through the example, let me point out two important consequences for root loci which follow from the general form of the characteristic equation

$$G(s)H(s) = -1/K$$

(1) As $K \to 0$, $1/K \to \infty$. So when K is small, the characteristic equation can only be satisfied where $|G(s)H(s)|$ is very large – in other words, near an open-loop pole. This means that **a branch of a root locus always starts at an open-loop pole location**.

If $G(s)H(s)$ has more poles than zeros the order of the denominator is greater than that of the numerator. So $G(s)H(s) \to 0$ as $s \to \infty$. If $G(s)H(s)$ has n poles and m zeros then the function is often said to have $n - m$ zeros at infinity.

(2) As $K \to \infty$, $1/K \to 0$. So when K is very large, the characteristic equation can only be satisfied where $|G(s)H(s)|$ is very small. This condition occurs either at an open-loop zero location or, if there are more poles than zeros in the open-loop transfer function, when s is very large. **Root-locus branches therefore end at an open-loop zero, or lead indefinitely towards the edge of the s-plane**.

Consider now the system illustrated in Fig. 6.15, which shows a proportional loop for controlling the angle of one axis of an industrial robot arm. Suppose, as usual that the controller gain is increased from a small value. How can we approach the problem of sketching the root locus in a methodical way?

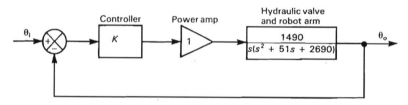

Fig. 6.15 A robot arm position control system.

Discussion

The open-loop poles are located at $s = 0$ and $s = -25.5 \pm j45.2$ (to three significant figures). There are no open-loop zeros. Using the earlier result for real axis branches of a root locus, it is clear that the whole of the negative real axis is part of the locus, since it lies to the left of one real pole. So the first step is to fill in the whole of the negative real axis as a branch of the locus, as shown in Fig. 6.16(a).

We also know that branches must start at the complex pole positions and eventually lead towards the edge of the s-plane (there are no finite open-loop zeros, but three 'at infinity'). What is the general behaviour of such branches as $K \to \infty$?

The answer is provided by 'standing back' from the s-plane, as illustrated in Fig. 6.16(b). At distances far enough from the pole cluster, the phase contours of the three poles will appear to originate in virtually the same place. Under these circumstances the individual contours are almost exactly superimposed. Now, the contours for *two* superimposed poles were shown in Fig. 6.9(b), where the two $\pm 180°$ contours left the double pole in opposite directions. Those for three superimposed poles can be obtained in a similar manner, but this time there will be *three* contours representing odd multiples of $\pm 180°$, spaced at 120° intervals. For the robot arm system, therefore, the root locus branches will follow this 120° spacing for very large K, as illustrated in Fig. 6.16(b). These lines represent the root locus branches as $K \to \infty$, and are called *asymptotes* in the root locus literature.

For large values of K, then, the root locus branches radiate symmetrically about the real axis with 120° between asymptotes. (Root loci must always be symmetrical about the real axis, since they represent closed-loop pole positions. These, by definition, must be real or occur in complex conjugate pairs.) We now need a little more information about behaviour closer to the pole cluster. For this I shall simply quote, without proof, a (reasonable enough) mathematical result:

The asymptotes intersect on the real axis at the centre of gravity s_g, defined by

$$s_g = \frac{\Sigma \text{ pole locations} - \Sigma \text{ zero locations}}{n - m}$$

where n is the number of poles and m is the number of zeros.

In this case, then we have

$$s_g = \frac{(-25.5 + j45.2) + (-25.2 - j45.2) + 0}{3}$$

$$= -17$$

Hence we can draw in the asymptotes as shown in Fig. 6.16(c).

We already have enough detail to infer some broad features of how the closed-loop response varies as gain is increased. For low gain there will be a real closed-loop pole near the origin dominating the response. As gain is increased this will move to the left along the real axis. However, if the complex closed-loop poles are to approach the complex asymptotes for large K, they must first move closer to, and eventually cross, the $j\omega$-axis. These complex poles will therefore first become dominant, giving increasingly noticeable oscillatory components in the transient response, and eventually move into the right half plane for sufficiently high gain, indicating instability.

To obtain more detailed information about the locus, the general phase condition can be applied to test points in the plane. For example, it is often useful to know the direction in which root-loci branches leave a complex open-loop pole. As always, the locus near to the complex pole must correspond to a total phase angle equal to an odd multiple of ±180°. Let us suppose that in this case a branch of the locus leaves pole P at an angle θ as shown in Fig. 6.16(d). At points very near to P the poles at $s = 0$ and $s = -25.5 - j45.2$ contribute $-120°$ and $-90°$, respectively, giving a total of $-210°$. If the root locus is to correspond to $-180°$ and hence satisfy the phase condition, the open-loop pole at P must itself contribute $+30°$ along the locus as it leaves the pole.

The $+30°$ contour from a pole corresponds to an angle of 30° *below* the horizontal (not above as was originally assumed in Fig. 6.16(d)). Using this final piece of information the complete root locus of the system can be filled in by symmetry: it is shown in Fig. 6.17.

Strictly speaking, an asymptote is approached but never actually reached. However, in the root-locus context, the term tends to be used irrespective of whether a branch of the locus follows such a line exactly, or simply approaches it asymptotically in the true mathematical sense. Note that there is no contradiction between the labelling of the real axis phase at $-\infty$ as ±540° in Fig. 6.16(b), and the earlier real axis result which implies a phase of ±180° to the left of one real pole. Figure 6.16(b) represents the situation for indefinitely large K, and is a limit condition applying strictly only at $K = \infty$. The important point is that the entire negative real axis is a branch of the root locus.

Note that the imaginary parts of the pole positions cancel out. Only real parts of pole or zero locations need to be included explicitly in this expression.

Refer to Fig. 6.7 if necessary.

Although the individual steps in sketching a root locus like this are all reasonably straightforward, it takes a great deal of practice to carry them out quickly and confidently. You should not expect to be able to sketch accurate root loci simply on the basis of reading this section.

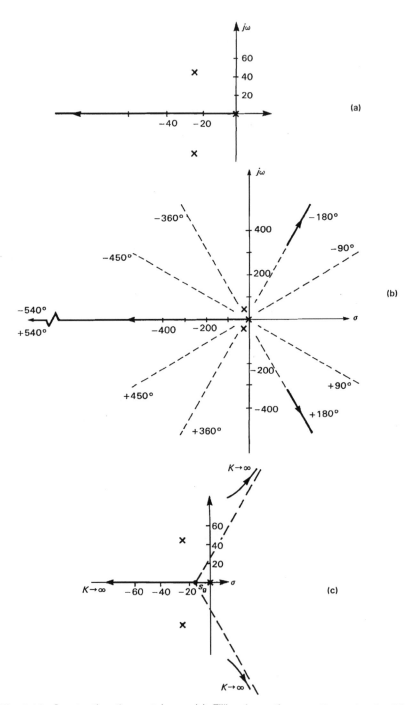

Fig. 6.16 Constructing the root locus. (a) Filling in sections on the real axis. (b) Asymptotes at infinity. (c) Locating the 'centre of gravity'.

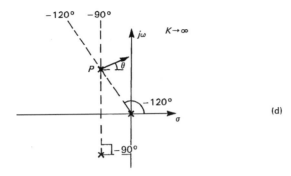

(d)

Fig. 6.16 Constructing the root locus. (d) Calculating the 'angle of departure'.

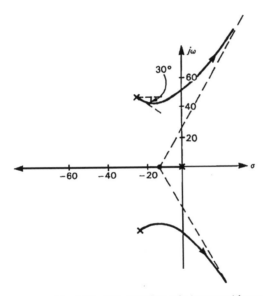

Fig. 6.17 The complete robot arm root locus.

A general procedure

Let me now suggest a step-by-step procedure for sketching a simple root locus.

(1) Plot the pole zero diagram and fill in sections of the locus along the real axis to the left of odd numbers of real poles plus zeros.

(2) If there are n poles and m zeros, calculate the centre of gravity at

$$s_g = \frac{\Sigma \text{ pole positions} - \Sigma \text{ zero positions}}{n - m}$$

125

(3) Sketch in $n - m$ radial asymptotes from the centre of gravity, symmetrically disposed about the real axis. (The angle between them will be $360°/(n - m)$, with the first lying at $180°/(n - m)$ to the real axis.)

(4) If necessary, use a consideration of local phase conditions to calculate the angles at which branches leave complex open-loop pole positions, or arrive at complex zeros. (Remember that the phase contribution of a zero is of an opposite sign to that of a pole.)

(5) Use symmetry arguments, and a knowledge of other, similar, root loci, to complete the sketch.

Root locus in analysis and design

Limitations of space mean that it is impossible here to give more than a flavour of how root loci can be used for system analysis and design. Rather than give a cursory treatment of a number of examples I have chosen to discuss one particular aspect of control system design in some detail, using simple root loci to illuminate system behaviour.

In many control systems oscillatory behaviour needs to be kept in check or even eliminated completely. Control of vehicle dynamics and certain servosystems are classic illustrations of this requirement. Several of the root loci of the previous sections have already indicated how increasing controller gain often leads to increasingly oscillatory closed-loop behaviour: the robot arm was a good illustration. In this section I shall concentrate on a simple model of an extreme example of this type of behaviour: a position control system designed to keep a communications satellite pointing towards the Earth with little or no steady-state error.

Figure 6.18 shows a basic proportional control loop for such a system, where the satellite's dynamics about one axis of rotation are modelled by the transfer function

$$\theta_o(s)/V(s) = K_s/s^2$$

Here θ_o is the angle of orientation, V is the voltage actuating signal to a positioning mechanism, and K_s is a constant depending on the drive characteristics and the moment of inertia (second moment of mass) of the satellite. The details of the positioning mechanism need not concern us here.

The open-loop transfer function of the system of Fig. 6.18 has a double pole at the origin and no zeros. The root locus of such a system has already been

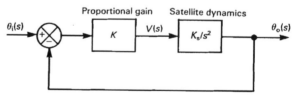

Fig. 6.18   A simplified satellite position control system.

Fig. 6.19   Satellite system root locus.

considered in Example 6.1: it is as repeated in Fig. 6.19. For positive controller gain the closed-loop poles are constrained to the $j\omega$-axis, being located where $s = \pm j\sqrt{K}$. In the configuration of Fig. 6.18, therefore, the closed-loop step response is a sustained oscillation at the natural frequency, as shown in Fig. 6.20: in terms of the standard second-order model, the system has a damping ratio of zero. No value of gain can give a suitable closed-loop response, since even the slightest disturbance to the system would result in undamped oscillations. Changing controller gain merely alters the natural frequency, not the damping ratio. Clearly, an important part of the control system in this example is to introduce damping so that the system settles to a steady-state output position.

Velocity feedback

A closer consideration of the step response of Fig. 6.20 suggests a way of generating appropriate control action. Because of the nature of proportional control, the same control action is generated at point P_1 as at P_2, the error is the same in each case. But, at P_2, the satellite is moving *towards* the desired position, while at P_1 it is moving *away*. It would therefore seem reasonable to reduce the control action at P_2 in comparison with that at P_1. Hopefully this would have the desirable effect of reducing the size of successive overshoots and eventually stabilize the response.

One way in which this can be done is illustrated in Fig. 6.21. Instead of feeding back simply the output position, the complete feedback signal consists of the

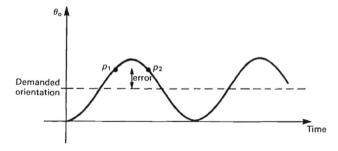

Fig. 6.20   Satellite system step response.

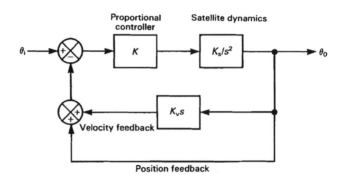

Fig. 6.21 Velocity feedback.

output plus a value proportional to the angular *velocity* of the satellite, represented
by the block with the transfer function $K_v s$. The velocity can be measured directly
by a sensor such as a rate gyro, or derived by differentiating the output of the
position sensor itself.

Then, while the satellite is moving away from the desired position, as at point
P_1, the value of the velocity $d\theta_o/dt$ is positive, and so the feedback signal, and
hence the control action, is increased. When the error is the same but the output
is returning towards the desired position, as at point P_2, the velocity $d\theta_o/dt$ is
negative, and the feedback signal and control action are decreased. Intuitively it
would seem that this additional *velocity feedback* should have the desired effect.

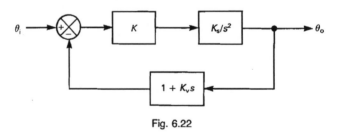

Fig. 6.22

The suggested control scheme can be redrawn as Fig. 6.22, where the two
feedback terms are now combined as a single transfer function. Now that the
system is drawn in this form we can use the root-locus technique to analyse
closed-loop behaviour. The open-loop transfer function of the system with velocity
feedback becomes

$$K(K_s/s^2)(1 + K_v s)$$

and the corresponding pole–zero plot is shown in Fig. 6.23. As can be seen, the
overall effect of the velocity feedback is to contribute a zero to the complete
open-loop transfer function at $s = -1/K_v$.

Now, since a branch of the root locus must end at the open-loop zero, the

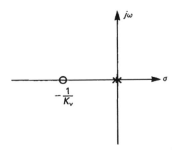

Fig. 6.23 Satellite system pole–zero plot with velocity feedback.

closed-loop poles are no longer constrained to the $j\omega$-axis. Figure 6.24 shows the new root locus. This is a typical example of how experience is needed to sketch root loci: the precise form of the figure cannot be deduced immediately from the guidelines presented earlier. Note, however, the following features, which do follow directly:

(a) the real axis to the left of the zero must be part of the locus (a total of three poles plus zeros to the right);
(b) because there is one more pole than zero only one branch of the locus tends to infinity, and if there is only one branch which tends to infinity it *must* lie along the negative real axis;
(c) the zero contributes zero phase at the origin, so the two root-locus branches leave the open-loop double pole in exactly the same direction as they did in Fig. 6.19.

The overwhelming impression of the plot is that the zero attracts a branch of the locus to itself. It may help to imagine open-loop poles as 'sources' of the locus branches, and open-loop zeros – or the edge of the s-plane – as 'sinks'.

In fact, this is a useful general way of visualizing the effect of zeros. Left half plane zeros have a stabilizing effect, since they attract loci branches towards a stable region of the plane. This is in clear contrast to the effect of additional poles

Only one branch of a locus can start at a single pole or end at a single zero. However, branches may coalesce (at points corresponding to double closed-loop poles) and then separate again, as in this case on the real axis to the left of the zero.

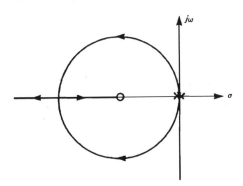

Fig. 6.24 Root locus with velocity feedback.

129

which, as noted in the context of an earlier example, tend to push branches of the locus towards the right half plane.

By altering the velocity feedback constant K_v, and hence the position of the zero, the precise form of the locus of Fig. 6.24 can be shaped so as to pass through appropriate locations for the closed-loop poles. Then, the proportional gain K can be adjusted as usual to locate the closed-loop poles at the desired positions on the locus.

It is instructive to analyse this simple example analytically. The closed-loop transfer function is given by

$$\frac{\theta_o(s)}{\theta_i(s)} = \frac{KK_s/s^2}{1 + (KK_s/s^2)(1 + K_v s)}$$

$$= \frac{KK_s}{s^2 + KK_s K_v s + KK_s}$$

This is a standard second-order lag transfer function with an undamped natural frequency given by

$$\omega_n^2 = KK_s$$

and damping ratio given by

$$2\omega_n \zeta = KK_s K_v$$

There is no zero in the closed-loop transfer function. The open-loop zero was contributed by the term in the feedback path; this term does not appear in the numerator of the closed-loop transfer function.

The two independent constants K and K_v allow the damping ratio and natural frequency of the closed-loop system to be chosen independently, within the range of validity of the model. Note that *increasing* the amount of velocity feedback (increasing K_v) for a given value of controller gain *increases* the damping ratio. This means that for a given controller gain, and hence ω_n, the oscillations in the step response die away faster as K_v is increased.

Worked Example 6.4 Assume for simplicity that K_s is equal to 1 in some appropriate units, and that the position control specification of the satellite can be satisfied with a damping ratio of 0.6 and an undamped natural frequency of $2\,\mathrm{rad\,s^{-1}}$. Choose appropriate values of K and K_v to satisfy these requirements. How long would the system then take to settle to within 2% of its steady-state value in response to a step input?

Solution If $K_s = 1$, then the closed-loop transfer function is

$$\frac{\theta_o(s)}{\theta_i(s)} = \frac{K}{s^2 + KK_v s = K}$$

Hence, comparing with the standard second-order form, we have

$$\omega_n^2 = K$$

and

$$2\zeta\omega_n = KK_v$$

To obtain $\omega_n = 2$, therefore, K needs to be set to 4, and to obtain a damping ratio of 0.6

$$2 \times 0.6 \times 2 = 4 \times K_v$$

Hence $K_v = 0.6$.

The settling time of a standard second-order step response to within 2% was given in Chapter 4 as

$$t_s \approx 4/\zeta\omega_n$$

In this case, then, $t_s \approx 4/1.2 \approx 3.3\,\mathrm{s}$.

Velocity feedback can also be used in other circumstances. Figure 6.25(a) shows the root locus of a unity feedback position control system like that of Fig. 6.3. Suppose that the specification implies that the closed-loop poles should be located at the positions indicated. Again the root locus can be changed to pass through these locations by means of appropriate velocity feedback, as illustrated by Fig. 6.25(b). The same technique can even be used when the plant is *unstable* in the absence of any control system, as shown in Fig. 6.25(c). In fact, velocity feedback is also useful for increasing the damping of oscillatory higher-order systems, since in many such cases the open-loop zero can be positioned so as to attract the dominant part of the root locus into a region of the s-plane corresponding to less oscillatory behaviour.

The precise way in which velocity feedback is implemented varies from system to system. For example, Fig. 6.26 shows a common alternative implementation where the velocity feedback is incorporated *after* the proportional gain, while Fig. 6.27 shows a form where the amounts of both position and velocity feedback can

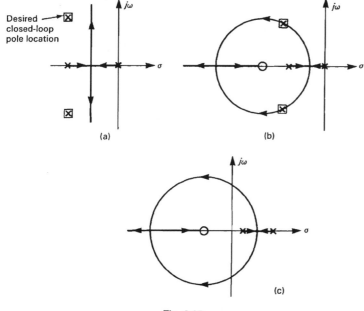

The term velocity, or rate, feedback is often used for any system with feedback of the derivative of the controlled variable, whether or not the latter has the dimensions of position.

Fig. 6.25

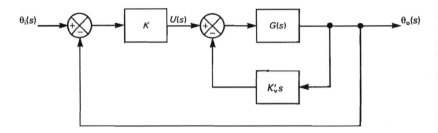

Fig. 6.26

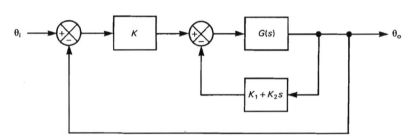

Fig. 6.27

be adjusted independently. Note that the effect of the velocity feedback in Fig. 6.26 is equivalent to that of Fig. 6.21 providing that $K'_v = KK_v$. This can be deduced directly from the two block diagrams: adding in a term K_vs *ahead* of a proportional element of gain K is equivalent to adding in a term $K'_vs = KK_vs$ *after* the gain.

Derivative action in the controller

Effects similar to those of velocity feedback can be achieved instead by including derivative action in the controller itself. In the $P + D$ *controller*, the control action $u(t)$ generated in response to an error signal $e(t)$ is

$$u(t) = K(e + T_d \, de/dt)$$

where T_d is known as the *derivative time*.

Although it is the error signal, rather than the feedback signal, which is differentiated, the result is still to contribute a zero to the open-loop transfer function, as can be seen from the controller transfer function

$$U(s)/E(s) = K(1 + T_ds)$$

which possesses a zero at $s = -1/T_d$.

Figure 6.28 shows the satellite example with a P + D controller in place of velocity feedback. Providing that $T_d = K_v$, this system, too, has the root locus of Fig. 6.23 as the proportional gain K is increased. In the case of Fig. 6.28, however (assuming again that $K_s = 1$ for simplicity), the closed-loop transfer function is

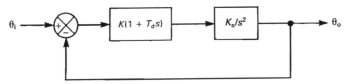

Fig. 6.28 Proportional + derivative control of satellite position.

$$\frac{\theta_o(s)}{\theta_i(s)} = \frac{K(1 + T_d s)}{s^2 + KT_d s + K}$$

Choosing $K = 4$, and $T_d = 0.6$, to correspond to Worked Example 6.4, results in an identical characteristic equation

$$s^2 + 2.4s + 4 = 0$$

and it is tempting to assume that the closed-loop behaviour would be identical. However, in the P + D case, the closed-loop transfer function *includes* the zero introduced by the derivative action, whereas in the velocity feedback case the zero occurs in the *open-loop* but not the closed-loop transfer function. The effect of this closed-loop zero with P + D control is similar to that described at the end of Chapter 5: the rise time is decreased, and there is additional overshoot. Step responses of the two cases are compared in Fig. 6.29. So, for example, if the satellite specification had restricted overshoot to 10%, say, then the velocity feedback design may have been satisfactory, but not the P + D design.

Once again, however, there are the usual trade-offs. The presence of the zero in the closed-loop transfer function can be advantageous in some circumstances, by providing greater initial control action and speeding up the response in comparison with velocity feedback. In other circumstances these very characteristics can be a problem: too much control action is produced initially, and undesirable overshoot occurs in response to a step change in input. Whether to use velocity feedback or the derivative of the error (or, indeed, some other option) depends very much on the circumstances and the system specification. The availability of suitable signals in the system for velocity feedback is also an important factor influencing the decision: in the case of mechanical servos and vehicle control, it is

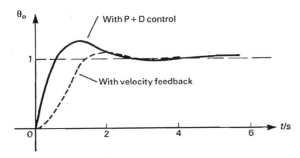

Fig. 6.29 Closed-loop step response of the satellite position control system.

often easy to add a velocity transducer of some sort, making velocity feedback an obvious candidate – although, here again, the benefits must be weighed against the expense of an additional sensor.

In the context of derivative action it should be noted that the performance of any control system in which a signal is *differentiated* (in contrast to *measuring an* additional velocity-type variable) may suffer because of the accompanying increase in noise. This is true irrespective of whether the differentiation takes place in the controller or in the feedback path. High frequency noise by definition changes rapidly, and therefore differentiating a signal contaminated by such noise can introduce a large unwanted element into a control signal. This in turn can lead to unnecessary control action and excessive wear in system components. Such noise problems can be partially overcome by using approximations to derivative action or by filtering differentiated signals appropriately.

This aspect will be taken up again later.

The above comparison of P + D and velocity feedback illustrates the danger of concentrating on the roots of the characteristic equation to the exclusion of every thing else, which is one of the drawbacks of the root-locus technique. In root locus design, the *complete* closed-loop transfer function must always be taken into account when predicting overall system behaviour. This is necessary both to check for the presence of closed-loop zeros (which may or may not feature on the *open loop* pole–zero plot) and, in higher-order systems, to ensure that poles believed to be dominant are in fact so. Effective root-locus design for higher-order systems therefore requires computer tools to enable such checks and simulations to be carried out fairly easily.

It must also be emphasized once again that manipulating models, as has been done throughout this chapter, is only a means to an end: namely, a final system design which performs as required. Designing for a particular value of dominant ζ or ω_n, for example, is only a convenient way of expressing certain system constraints and specifications. And just because a design model predicts particular values of such parameters it does not mean that these values will necessarily be achieved in practice, or even that the values chosen are the best way of satisfying the specification. Furthermore, the designer always needs to take into account those aspects of system reality which may be concealed by the beguiling ideal-izations of the mathematical model. The nature of intermediate signals in a control loop – the output of the controller, for example, which determines the demands on valves and actuators – is not revealed explicitly by our convenient pole–zero plots and transfer functions. Yet such aspects can profoundly influence the success or failure of a control scheme, and need to be carefully examined before implementing a final design.

Summary

This chapter has outlined how the root-locus technique can be used to determine the way closed-loop pole positions vary as loop gain is increased from a small value. Such root-locus plots can be derived from a knowledge of the open-loop pole–zero configuration alone. The root locus can be shaped by introducing additional poles and zeros into the open-loop transfer function. Additional poles tend to push branches of the locus further to the right, while zeros attract branches of the locus. If positioned appropriately, therefore, zeros can increase damping and stabilize system closed-loop response. Two common ways of introducing such zeros were presented: velocity feedback, and the proportional + derivative (P + D) controller.

One disadvantage of the root-locus technique is that it only gives information about closed-loop poles. A separate check always needs to be made that any closed-loop zeros do not affect system behaviour detrimentally, and also that any closed-loop poles assumed to be dominant are in fact so. Such checks can involve time-consuming calculation, particularly when using the root-locus technique for higher-order systems. In such cases computer tools are invaluable: for generating the initial root loci; to assist in the selection of controller parameters; and to simulate proposed design solutions.

Problems

6.1 A P + I controller with the transfer function

$$C(s) = K(1 + 1/T_i s)$$

is to be used in a unity feedback configuration to control a process modelled by the transfer function

$$G(s) = 1/(1 + 5s)$$

Sketch the root locus as controller gain K is increased from a small value for $T_i = $ (a) 10s and (b) 3s. Noting that *reducing* T_i corresponds to *increasing* the amount of integral action, comment on the significance of the differences in the two loci.

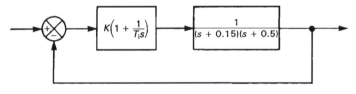

Fig. 6.30

6.2 Figure 6.31 shows the root-locus of the system of Fig. 6.30 for $T_i = 10$s. Locate the new asymptotes if T_i is decreased to (a) 5s; and (b) 1s, and sketch the complete loci for these cases. What is the minimum value of T_i such that the model predicts closed-loop stability for all positive values of controller gain?

6.3 A hydraulic system used for position control has the process transfer function

$$G(s) = \frac{1500}{s(s^2 + 140s + 4890)}$$

(a) Sketch the root locus of the corresponding unity feedback system with a proportional controller.

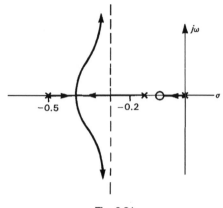

Fig. 6.31

(b) Using your root-locus sketch, estimate the value of proportional gain required to give a dominant closed-loop real pole corresponding to a time constant of 0.1 s.

6.4 (a) Sketch the root locus of the system of Fig. 6.6.
   (b) Another system can be represented by the same block diagram, but this time $\tau = 2\,\text{s}$, $\omega_n = 10\,\text{rad}\,\text{s}^{-1}$ and $\zeta = 0.7$. Sketch the new root locus.
   (*Hint*: In each case consider carefully the angle at which the loci branches leave the complex open-loop poles.)

6.5 The root loci of the systems of Figs 6.32 and 6.13 are identical. Derive the closed-loop transfer functions of each for a general controller gain K and comment on the difference.

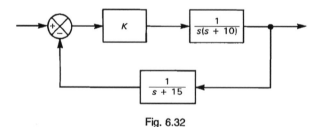

Fig. 6.32

6.6 The system controlling the position of a workpiece can be modelled as shown in Fig. 6.33.
   (a) Derive the closed-loop transfer function $\theta_o(s)/U(s)$ of the inner loop in terms of K_v.

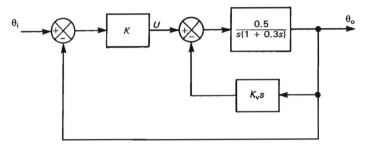

Fig. 6.33

(b) Hence obtain the closed-loop transfer function of the complete system in terms of K and K_v.

(c) Suggest appropriate values of K and K_v to obtain a closed-loop response with damping ratio 0.7 and ω_n of $3 \, \text{rad s}^{-1}$.

7 Steady-state performance

Objectives
  ☐ To discuss qualitatively some important aspects of steady-state performance.
  ☐ To introduce the concept of tracking as part of a system specification.
  ☐ To show how transform equations can provide quantitative estimates of steady-state errors.

The previous two chapters have concentrated on system transient behaviour. In this chapter we return to the steady-state performance of a control system – that is, the behaviour of the system once the transient response has died away. This definition of the steady-state condition, given a formal framework in Chapter 5, presupposes that the system is stable – otherwise there would not be a transient response which dies away with time. As has been seen before, it is a definition general enough to include steady states where the input and output are changing with time. The salient feature of such steady-state conditions is that the input and output have the same *general forms*. We have already seen a number of examples of this general principle: the constant input/constant output condition for step response steady state; the sinusoidal input/sinusoidal output of frequency response testing; and a ramp input/ramp output steady state in one of the worked examples of Chapter 5.

So far we have considered in detail the steady-state error of a closed-loop system in the context of one type of input only: response to a step change in input or a steady disturbance. This aspect was first discussed in Chapter 2 in terms of steady-state gain, and then taken up again in Chapter 4 in the context of frequency response and integral action. Yet steady-state performance may be specified for other types of input. Perhaps the most common is to specify the permissible steady-state error when the input and output of a control system are both changing at a constant rate. For example, in machining operations, the error between the demanded and actual position of a machine tool or workpiece may need to be kept below a specified value, or even eliminated altogether, while the tool or workpiece carrier is moving at a given rate. The ability or otherwise of a controlled variable to *track* an input changing at a given rate is therefore often a crucial aspect of steady-state performance.

> Strictly speaking, we also need to include a number of special cases, such as the possibility of a zero steady-state output for a given finite input. For example, the steady-state output of a high-pass filter in response to a step input is zero, since the filter eliminates the zero-frequency 'd.c. component' of the input. Input and output still possess the same general form, however, if we consider the latter to be a constant of zero magnitude.

An intuitive approach

Although mathematical modelling is indispensable for a quantitative analysis of steady-state behaviour, a qualitative or intuitive treatment is also instructive. To begin with, I shall discuss steady-state performance in such a way, before turning to transform equations to provide numerical estimates for particular situations.

Even an intuitive approach needs some sort of system model, however. To simplify matters, let us assume that the process to be controlled has the following

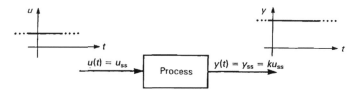

Fig. 7.1   Steady-state step response.

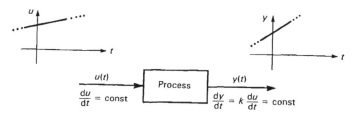

Fig. 7.2   Steady-state ramp response.

steady-state input–output characteristics, which are illustrated in Figs 7.1 and 7.2:

(1) If the input $u(t)$ to the process in the steady state is a constant u_{ss}, then the output is a constant $y_{ss} = ku_{ss}$ where k is the steady-state gain.
(2) If the input to the process is changing at a constant rate in the steady state, then the output also changes at a constant rate such that

$$dy/dt = k\, du/dt = \text{constant}$$

where k is again the steady-state gain.

These two assumptions fit in well with practical experience of many real systems. What they imply is that to maintain the output of the process at a constant, non-zero level requires a constant, non-zero input. Similarly, to keep the output increasing at a constant rate needs a constantly increasing input. Of course, there are systems which do not behave like this (more of this later), but the analysis can easily be adapted to suit them.

Note that these two assumptions or conditions are consistent with common dynamic process models. Suppose, for example, that the process is modelled as a first-order lag, so that

$$\tau \dot{y} + y = ku$$

Then, under steady-state condition (1) above, $u(t) = u_{ss}$; $y(t) = y_{ss}$; and $\dot{y} = 0$. Then we have $y_{ss} = ku_{ss}$ as before. Condition (2) also follows simply. Differentiating the system equation gives

$$\tau \ddot{y} + \dot{y} = k\dot{u}$$

If, in the steady state, the input and output are both changing at a constant rate we have $\ddot{y} = 0$ and $\dot{y} = k\dot{u}$ as before.

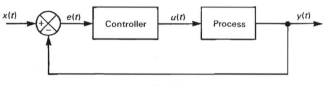

Fig. 7.3

Suppose now that the output of a process with these steady-state characteristics is to be controlled using the simple feedback loop of Fig. 7.3, where we assume throughout the discussion that the closed loop is stable. In Fig. 7.3, $x(t)$ represents the desired value of the controlled variable and $y(t)$ the actual value. In such a unity feedback configuration the output of the differencing element, $e(t)$ is a direct measure of the error – that is, the difference between the desired and actual values of the controlled variable.

Care needs to be taken when reading accounts of steady-state error. Authors sometimes define error more generally as the difference between any reference input and the feedback signal. For non-unity feedback, or a reference input which is not identical to the demanded value of the controlled variable, this convention does not correspond to that adopted here.

Let us first revise system steady-state properties under condition (1) above. That is, we require the process to settle down to a constant output after a change in input to a new, constant value. If in the steady state the output of the process is y_{ss}, then by assumption (1) the process input must be a steady $u_{ss} = y_{ss}/k$. This steady control action must be provided by the controller. In Chapter 4 it was argued that if a finite steady-state error e_{ss} can be tolerated, then a proportional gain alone may be feasible as a controller. The error signal provides the input needed by the gain element to generate the appropriate control action. If e_{ss} is to be reduced to zero, on the other hand, then the controller must be able to supply the finite output u_{ss} even in the absence of any error input. This requires *integral action* in the controller: the output of an integrator, as was illustrated in Fig. 4.15, can remain at a finite value even in the absence of any input.

The integrator output may be finite or zero for a zero input signal, depending on the previous history of the input. Note that this general property of an integrator ties in with its marginal stability. In contrast, the output of an *absolutely* stable system (all poles in the left half plane, as has been assumed for the process) decays to zero when the input is removed – that is, the system returns to the quiescent state.

So far the argument should have been familiar from Chapter 4. Let us now turn to the tracking problem, and suggest conditions under which the output of a closed-loop system can follow the input with finite or no steady-state error. Applying the same general reasoning as before we suppose that in Fig. 7.3 the output $y(t)$ must increase at a constant rate in the steady state, in response to a steadily increasing demand value $x(t)$. The first point to note is that the slope of the input x must equal the slope of the output y. Otherwise, the error e will increase indefinitely with time, behaviour unacceptable in a control system whose output is required to track a steadily changing input.

Let us assume first that the specification permits a constant steady-state error e_{ss}. This state of affairs is illustrated in Fig. 7.4. The output y has the same slope as the input, but lags behind it by a constant value e_{ss}. But we have assumed that to drive the process in this way needs a constantly increasing control action u (assumption (2)).

In fact, the slopes of u and y are related by the expression, derived earlier, $dy/dt = k\,du/dt$.

To give a finite steady-state error, therefore, the controller must be able to produce a steadily increasing control action in response to a *constant* input e_{ss}. What sort of device must such a controller be? Clearly, a proportional gain is inadequate; in fact, the controller must again provide integral action. A constant positive input to an integrator produces a continuously increasing output with a constant slope, just as is required to drive the plant. (Recall the integrator step response of Chapter 3.)

Of course, this does not mean that the controller is simply an integrator,

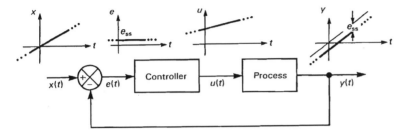

Fig. 7.4 Tracking with finite steady-state error.

although there are systems where this may be the case. The P + I controller, as noted in Chapter 4, also possesses the desirable integral action, and gives the designer additional flexibility in the independent choice of gain and integral time. Recalling that the controller behaviour may be expressed as

$$u(t) = K\left[e(t) + \frac{1}{T_i}\int e(t)\,dt\right]$$

we can draw the controller block diagram as shown in Fig. 7.5. The output of the integrator block ensures that the complete controller output is of the required form in the steady state.

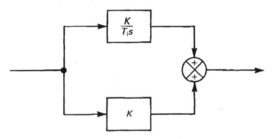

Fig. 7.5 A P + I controller.

Finally, let us consider the problem of eliminating steady-state error completely under tracking conditions (Fig. 7.6). Then the controller must continue to supply a ramp-type control action *even after the steady-state error e_{ss} has been reduced to*

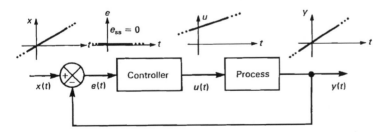

Fig. 7.6 Tracking with zero steady-state error.

141

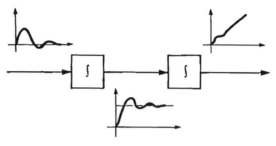

Fig. 7.7 Double integral action.

zero by the action of the control loop. To do this requires not one but *two* cascaded integrators in the controller, as illustrated in Fig. 7.7. A single integrator can provide a steady, finite output with zero input, while the second integrator can convert this into the required ramp.

Worked Example 7.1   Figure 7.8 shows a position control system. Give a qualitative explanation of why single integral action in the controller eliminates steady-state error when tracking an input changing at a constant rate. Assume that the controller includes appropriate elements to stabilize the closed-loop system.

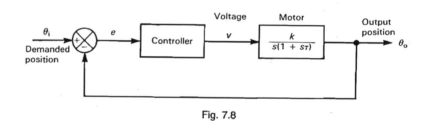

Fig. 7.8

Solution   This may be approached in two ways. The first is to reason as follows. The motor does not satisfy the assumptions made about the process in the preceding general discussion. Its own integrating action means that its output (position) increases indefinitely in response to a constant input (voltage). Hence to generate a ramp position output in the absence of steady-state error requires a constant control action from the controller, rather than a steadily increasing signal. A controller with a single integrating action is therefore sufficient.

Alternatively, we note that the behaviour of the system of Fig. 7.8 is identical to that of Fig. 7.9. By 'detaching' the integrating action of the motor, the system has been converted to the form of the previous analysis. Hence it is clear that the actual controller only needs to supply a single integrator.

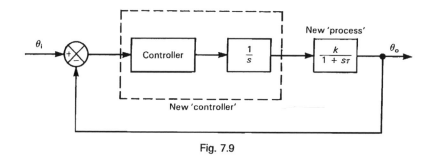

Fig. 7.9

The results of the preceding qualitative analysis of tracking can be summarized as follows:

(1) For a unity feedback closed-loop system to track, with finite steady-state error, an input changing at a constant rate, there must be a single integrator in the forward path – that is, the forward path transfer function must possess a single pole at the origin.
(2) For a unity feedback closed-loop system to track, with zero steady-state error, an input changing at a constant rate, there must be **two** integrators in the forward path – that is, the forward path transfer function must possess a **double** pole at the origin.

Similar results can be derived for steady-state error in response to disturbance inputs, rather than command inputs, the difference being that the relevant integrator(s) must be upstream of the point of entry of the disturbance. It would be unusual, however, to specify that a control system must eliminate steady-state error in response to a disturbance *changing* at a constant rate. On the other hand, specifying zero steady-state error in response to a *steady* disturbance is common, as was considered in Chapter 4 in the context of the P + I controller.

The transform approach

Let us now turn to a quantitative analysis of specific systems. Figure 7.10 shows the position control system of Fig. 7.8 with a proportional controller. Suppose that the steady-state position error is to be kept within a given limit when the input and output are changing at a constant velocity, Ω. The previous qualitative analysis suggests that tracking with constant steady-state position error is indeed possible for this system, since the forward path includes the integrating action of the motor itself.

Note that we are still considering *position error*, even though the input and output are changing with time. Position error under these circumstances is often (rather confusingly) referred to as **velocity lag.**

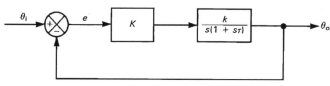

Fig. 7.10

A quantitative expression for the magnitude of the position error in these circumstances can be obtained easily from the transfer function relating the error e to the system output θ_o. From the block diagram we have

$$\frac{\theta_o(s)}{E(s)} = \frac{Kk}{s(1 + s\tau)}$$

Hence

$$s^2\tau\theta_o(s) + s\theta_o(s) = KkE(s)$$

Writing this transform equation as a differential equation gives

$$\tau\ddot{\theta}_o + \dot{\theta}_o = Kke$$

Now, in the steady state the output tracks the input at the same velocity but with a constant position error e_{ss}. The first derivative of θ_o is therefore constant, and equal to the velocity Ω, while the second (and higher) derivatives of θ_o are zero. We can therefore write down immediately an expression for the steady-state position error e_{ss}

$$\Omega = Kke_{ss}$$

or

$$e_{ss} = \Omega/Kk$$

From this expression it is clear that the magnitude of the position error when tracking at a constant velocity is reduced by increasing the controller gain.

Note that the intermediate step of writing out the differential equation in the above derivation is unnecessary. The expression for the steady-state error can be written down immediately from the transform equation by noting the correspondence between transform terms in s and s^2 and time domain terms in velocity and acceleration.

Worked Example 7.2   A position control system is modelled as shown in Fig. 7.11. The step response must not overshoot by more than 5%, and the system must be able to track an input changing at a constant velocity Ω with a position error whose magnitude is less than 0.1Ω. That is, if the input is changing at $1\,\mathrm{rad\,s^{-1}}$ then the steady-state position error should be less than $0.1\,\mathrm{rad}$. Can this specification be satisfied by a proportional controller?

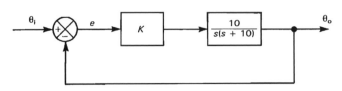

Fig. 7.11

Solution   The closed-loop transfer function can be derived as usual as

$$\theta_o(s)/\theta_i(s) = 10K/(s^2 + 10s + 10K)$$

This is in the familiar second-order form, and for an overshoot of less than 5% in the step response we require a damping ratio greater than 0.7. This gives a limit on maximum controller gain, since

$$2\zeta\omega_n = 10$$

and

$$\omega_n^2 = 10K$$

Hence with $\zeta = 0.7$, $\omega_n = 10/1.4 = 7.14$, and

$$10K = (7.14)^2 = 51.$$

This gives an upper limit of $K = 5.1$.

Consider now the tracking performance. From the block diagram we have

$$\theta_o(s)/E(s) = 10K/s(s + 10)$$

Hence

$$s^2\theta_o(s) + 10s\theta_o(s) = 10KE(s)$$

The term in s^2 corresponds to acceleration in the time domain, while the term in s corresponds to velocity. In the steady-state tracking condition the acceleration is zero, and the time domain equivalent to the above transform equation becomes

$$10\Omega = 10Ke_{ss}$$

where Ω is the rate of change of input and output position, and e_{ss} is the constant steady-state error. Hence

$$e_{ss} = \Omega/K$$

To meet the specification, $e_{ss} < 0.1\Omega$, requires $K > 10$, which is incompatible with the step response requirement of $K < 5.1$. In fact, for $K = 10$, the closed-loop damping ratio is around 0.5 and the step response overshoot about 15%.

A similar analysis is possible for systems with non-unity feedback, as illustrated in the following example. Note that in such cases the steady-state error must be calculated by taking into account the nature of the feedback path.

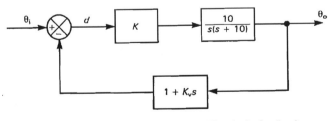

Fig. 7.12   A position control system with velocity feedback.

Velocity feedback is used in the position control system of Worked Example 7.2 in the form shown in Fig. 7.12. Derive, in terms of proportional gain K and velocity feedback constant K_v, an expression for the steady-state position error when tracking at a constant velocity Ω.

Solution Denote the output of the differencing element by $d(t)$, since with non-unity feedback this no longer corresponds to the error in the strict sense of the difference between desired and actual values of the controlled variable. Then we can write

$$\theta_o(s)/D(s) = 10K/s(s + 10)$$

or

$$s^2\theta_o(s) + 10s\theta_o(s) = 10KD(s)$$

Converting directly into a time domain steady-state expression we have

$$10\Omega = 10Kd_{ss}$$

Hence

$$d_{ss} = \Omega/K$$

This expression now needs to be related to the position error $e_{ss} = \theta_i - \theta_o$. From the block diagram the following equation holds in the time domain:

$$d_{ss} = \theta_i - (\theta_o + K_v\Omega)$$

and hence

$$\begin{aligned} e_{ss} = \theta_i - \theta_o &= d_{ss} + K_v\Omega \\ &= (\Omega/K) + K_v\Omega \end{aligned}$$

For a given proportional gain, therefore, the presence of velocity feedback *increases* the position error when tracking an input changing with constant velocity. The error increases as the velocity feedback constant K_v is increased.

It is interesting to compare this result with the corresponding one for the system of Fig. 7.13, where the velocity feedback has been replaced by derivative action in the controller. In this case the output of the differencing element directly represents system error again. In the steady state, with the output tracking the input with constant position error, the derivative action will not contribute to the control action. The steady-state analysis will therefore be entirely analogous to

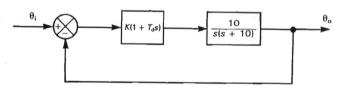

Fig. 7.13   A position control system with a P + D controller.

that for a proportional controller, and the steady-state position error will be the same, namely $e_{ss} = \Omega/K$. Derivative action in the controller, unlike velocity feedback, does not affect tracking performance for a given proportional gain.

Summary

A qualitative analysis of the steady-state performance of simple closed-loop systems leads to the following important conclusions.

(1) A single integrator in the forward path is necessary to eliminate steady-state error in response to a step change in input, or to give finite steady-state error when tracking an input changing at a constant rate.
(2) To eliminate steady-state error under tracking conditions needs *two* integrators in the forward path.

Transform equations may be used when a quantitative estimate of steady-state error is required. Such an analysis shows that:

(a) tracking error, if it exists, can be reduced by increasing controller gain; and
(b) the presence of velocity feedback increases such error.

Although this chapter has concentrated exclusively on steady-state performance, such analyses cannot be separated from consideration of transient response. For example, steady-state analysis assumes that the closed-loop system is stable, an assumption which must be checked in practice. Moreover, any measures to improve steady-state performance (such as increasing gain or incorporating integrators) also affect transient response. In the context of design, therefore, steady-state analysis must be carried out in conjunction with appropriate checks on stability and transient performance.

The treatment of steady-state error given in this chapter can be formalized in terms of so-called error constants for various types of system input (step, ramp, etc.). However, a direct analysis of the appropriate transform equation gives a clearer idea of what is going on in the system, so is the approach adopted here.

Problem

7.1 What is the steady-state position error of the control system of Problem 6.6, with parameters as selected in part (c), in response to
   (a) a unit step change in demanded position;
   (b) an input changing steadily at the rate of $1 \, \text{rad s}^{-1}$?

8 Controllers and compensators

Objectives

☐ To introduce three-term (PID) controllers as a combination of P + I and P + D types.

☐ To present empirical rules for tuning three-term controllers.

☐ To introduce phase compensation, and describe the steps in a frequency response approach to compensator design.

☐ To outline some aspects of cascade and feedforward control.

Alongside classical control theory is an approach usually termed 'modern' control, which was developed in the 1960s. Instead of analysing systems in terms of possibly high-order input–output transfer functions, modern control theory describes dynamic behaviour in terms of sets of first-order differential equations. These sets of equations relate so-called state variables (which do not necessarily correspond to system inputs or outputs in the sense used in this book), and are normally handled using matrix techniques.

Naturally, these remarks are broad generalizations, and will need to be treated with a degree of caution when dealing with specific systems.

The preceding chapters have introduced the fundamental modelling tools and building blocks of what is often called 'classical' control theory. By this is meant the frequency response and s-plane ideas which allow us to construct powerful conceptual models both of the dynamics of the systems to be controlled, and of the functional characteristics of the control elements necessary. So, for example, in Chapter 4 the idea was introduced of 'shaping' open-loop frequency response to give required closed-loop behaviour, and in Chapter 6 the idea of modifying the open-loop pole–zero plot with the same aim.

The basic control system building blocks used so far have been variable gains, integrators and differentiators. These elements have been characterized by their frequency response curves and their transfer functions, and in principle are all we need to design a controller for any linear system. In frequency domain terms, a suitable combination of these elements should allow the frequency response to be shaped successfully; in s-plane terms, the addition of sufficient poles and zeros in appropriate locations should result in a satisfactory closed-loop pole–zero configuration. Of course, compromises and trade-offs are always involved. Before proceeding further, therefore, it is useful to summarize some of these compromises in terms of the general effects on system behaviour of the control options introduced so far.

Steady-state performance

The steady-state performance of a system will usually be expressed in terms of the permissible error for particular types of input, most commonly step or ramp inputs. Steady-state analysis indicates that errors can be reduced by increasing controller gain, and eliminated (in theory at least) by ensuring the presence of sufficient integrators (depending on the type of input) in the forward path. For disturbance rejection the gain element or integrator(s) must be located upstream of the point of entry of the disturbance. Then appropriate control action can be generated without the disturbance being amplified or integrated by the controller.

Transient response

A system will normally be required to satisfy certain limits on transient behaviour, usually expressed in terms of rise time, settling time, bandwidth, ω_n and ζ (for a dominant second-order response), and so on. Increasing controller gain or including derivative action tends to speed up system response (in the sense of decreasing rise-time or increasing bandwidth), while adding integral action tends to have the reverse effect.

In s-plane terms, speed of response requirements will place a limit on how close dominant closed-loop poles should be to the $j\omega$-axis. Overshoot requirements will place a limit on ζ and hence on the damping angle.

Stability

Both the frequency response and root-locus techniques of earlier chapters indicate that increasing controller gain, or adding integrators to a loop, tends to reduce stability margins, possibly leading to oscillatory transient behaviour or even instability. Derivative action, on the other hand, can increase the damping of oscillatory systems and even stabilize unstable systems.

With these broad characteristics in mind, let us now turn to some control system options other than those considered in earlier chapters.

Three-term controllers

The compromises inherent in P + I control on the one hand and P + D control on the other mean that sometimes neither approach alone can simultaneously satisfy both the steady state and the transient response aspects of system specification. In such cases the logical strategy is to attempt to combine the best features of both. Thus we arrive at the *three-term* or *PID controller*, which produces an output $u(t)$ in response to an error $e(t)$ given by the equation

$$u(t) = K\left(e(t) + \frac{1}{T_i}\int e(t)\,\mathrm{d}t + T_d\frac{\mathrm{d}e}{\mathrm{d}t}\right)$$

The corresponding transfer function is:

$$U(s)/E(s) = K(1 + 1/sT_i + sT_d)$$
$$= \frac{KT_d(s^2 + s/T_d + 1/T_iT_d)}{s}$$

The PID controller therefore possesses a pole at the origin, like the P + I controller, but now *two* zeros, whose precise locations are determined by the values of the integral and derivative times. In theory, the positions of these two zeros, and the value of the proportional gain, can all be set independently, with the zeros being real or a conjugate complex pair as required. The designer therefore has sufficient additional flexibility to satisfy system specifications in certain circumstances where neither a P + I nor a P + D controller alone is adequate. The additional complexity, however, can make it difficult to decide exactly how to choose values of the three controller parameters.

One common approach to setting PID parameters is to choose $T_i = 4T_d$ so that the transfer function becomes

$$\frac{U(s)}{E(s)} = \frac{KT_d(s + 1/2T_d)^2}{s}$$

PID controllers are by far the commonest controller in industrial process control. It has been estimated, however, that in over 90% of these the derivative action is turned off!

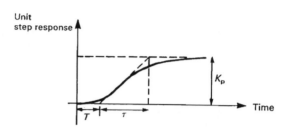

Fig. 8.1 A process step response or 'reaction curve'.

— that is, there is a double zero located at $s = -1/2T_d$. The most desirable position for the double zero can be determined using frequency response methods to satisfy gain and phase margins, or root-locus methods to position the closed-loop poles as desired. In addition, there are also various empirical rules, which have been found to give acceptable results for a wide range of plants.

Empirical methods of setting controller parameters

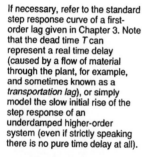

If necessary, refer to the standard step response curve of a first-order lag given in Chapter 3. Note that the dead time *T* can represent a real time delay (caused by a flow of material through the plant, for example, and sometimes known as a *transportation lag*), or simply model the slow initial rise of the step response of an underdamped higher-order system (even if strictly speaking there is no pure time delay at all).

The Ziegler and Nichols settings were designed primarily to reject *disturbances* in the type of plant not normally subjected to step changes in set point. So the highly oscillatory nature of the step response is not so significant as might appear.

In Chapter 3 it was pointed out that many practical processes exhibit a broadly s-shaped open-loop step response, as illustrated in Fig. 8.1. Such processes can often be modelled successfully by a process transfer function of the form

$K_p e^{-sT}/(1 + s\tau)$

The steady-state gain K_p, dead time T and time constant τ can be estimated from the open-loop step response curve. A tangent drawn to the point of maximum slope, as illustrated in Fig. 8.1, intersects the time axis and the line representing the steady-state value at times $t = T$ and $T + \tau$, respectively.

In 1942, Ziegler and Nichols suggested two empirical methods for setting controller parameters, based on an analysis of actual plant operating records. The aim was to produce a closed-loop step response of the form shown in Fig. 8.2, in which successive overshoots are reduced by a factor of four. The controller settings suggested by the first of these two methods assume a measured or simulated open-loop step response of the form of Fig. 8.1. Table 8.1 lists the recommended settings for such a process, expressed in terms of the parameters K_p, T and τ.

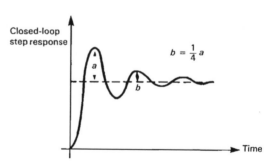

Fig. 8.2 Quarter amplitude response.

Table 8.1

Controller type	Gain K	Integral time T_i	Derivative time T_d
Proportional	$\dfrac{\tau}{T \times K_p}$	—	—
P + I	$\dfrac{0.9\tau}{T \times K_p}$	$3.3T$	—
P + I + D	$\dfrac{1.2\tau}{T \times K_p}$	$2T$	$0.5T$

Table 8.2

Controller type	Gain K	Integral time T_i	Derivative time T_d
Proportional	$0.5K_U$	—	—
P + I	$0.45K_U$	$T_U/1.2$	—
P + I + D	$0.6K_U$	$T_U/2$	$T_U/8$

In outline, this is usually done as follows. Start with a sufficiently low value of gain for the closed-loop system to be stable; subject the set point to a small step change from its operating point, and then return it to its original value. Increase the gain slightly and repeat as necessary. Initially any transient oscillations of the output will die away, but as the procedure is repeated with increased gain, the stability limit will be approached and the output will oscillate indefinitely.

The second method is applied to a process already under closed-loop control, rather than relying on the results of open-loop testing. First, any derivative or integral action is turned off, leaving a purely proportional controller. The proportional gain is then gradually increased from a low value to the point at which the system is on the verge of instability, with the output oscillating indefinitely. The recommended controller settings are based on these 'ultimate' values of gain, K_U, and period of sustained oscillation, T_U, as listed in Table 8.2.

Worked Example 8.1

What controller settings are suggested by the Ziegler and Nichols open-loop method for a P + I controller for the temperature control system of Chapter 4? (The process was modelled by a step response of the type shown in Fig. 8.1, with a dead time of 1 s, a time constant of 10 s, and a steady-state gain of 1.) Comment on any differences between the values suggested by Table 8.1 and those obtained in the design example of Chapter 4.

Solution In this example, $T = 1$, $\tau = 10$ and $K_p = 1$. Values of proportional gain K and integral time T_i follow immediately from the table, giving

$$K = 0.9 \times \tau/(T \times K_p) = 9$$

and

$$T_i = 1/0.3 \approx 3.3$$

The integral time is close to the final value selected in the design example of Chapter 4 (4 s) but the value suggested for the proportional gain is about 6 dB higher (9 as opposed to 5).

A brief reconsideration of the basis of the Ziegler and Nichols method indicates why this should be so. In the design example of Chapter 4, the aim was to restrict

Use the Nichols plots of Chapter 4 to compare the closed-loop step responses with proportional gains of 5 and 9.

overshoot to 25%. The 'quarter amplitude response' of Fig. 8.2, on the other hand, has a much greater initial overshoot. (In fact, for a second-order system the damping ratio corresponding to the latter response is around 0.2.) While the Ziegler and Nichols approach is still useful in practice, it may be desirable to reduce the suggested proportional gain somewhat, depending on the particular system and its specification. In fact, various detailed suggestions for modifying the original values of Table 8.1 have been made in the control engineering literature in any event, however, empirical rules can only offer a starting point for controller fine tuning.

Compensators

Compensators may be defined as components which are added to a control system in order to modify closed-loop performance. The same can of course be said about controllers, and indeed there is very little difference between a controller and a compensator from the design point of view. The two terms reflect differences in practice and hardware, rather than differences in design approach. Traditionally a controller is a stand-alone device offering a selection of specific control options such as proportional gain and integral and derivative action. Controllers normally include the differencing element, a number of inputs and outputs (suitable for the various signal levels which are standard in the process industry) and controls for adjusting parameters and set points. Compensators, on the other hand, consist solely of the elements for altering system dynamics and are often built as required out of individual components. As is the case with controllers electrical or electronic devices are most common, although mechanical or pneumatic compensators are also possible.

In theory at least, compensators can be added anywhere in a control system: in the forward path, in the feedback path, as an inner loop. Consideration here will be restricted to those in the forward path, as illustrated in Fig. 8.3; this is often known as **series compensation**.

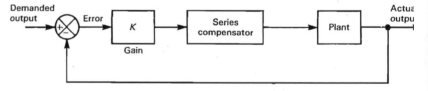

Fig. 8.3 Series compensation.

Lead compensation

Ideal P + D control, with perfect differentiation, cannot be realized in practice. Figure 8.4 shows the Bode plot of an ideal P + D controller. The $(1 + j\omega T_d)$ term in the numerator of the frequency response function implies an amplitude ratio which rises indefinitely (at 20 dB per decade) as frequency is increased. No real system can behave like this, however: at sufficiently high frequencies the amplitude

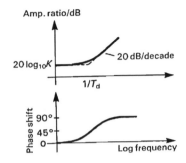

Fig. 8.4 Ideal P + D frequency response.

ratio of any physical device is bound to roll off. What matters in a control system, though, is the behaviour of the controller over the range of frequencies to which the plant can respond, and it is often possible to approximate derivative action fairly well over such a frequency range – for example, by using fast electronic circuits in the controller for a plant with comparatively long time constants. Another problem with derivative action, however, is susceptibility to noise, and again this can be appreciated from the Bode plot of Fig. 8.4. The effect of the continuously rising amplitude ratio is to amplify higher-frequency noise, and this can then contribute to the control action even when it does not properly reflect the behaviour of the plant. This is undesirable for a number of reasons. For example, it can cause excessive wear in those components which are driven by the controller output – the so-called *final control elements* such as valves and actuators – and it is generally pointless anyway to be taking control action which does not originate in significant plant behaviour! For such reasons, therefore, if P + D control is indicated (in order to increase closed-loop bandwidth or improve stability margins, for example), it may be more desirable to aim to implement a 'smoothed' version, in which the high-frequency components of the compensator output are reduced. This idea can be represented in general terms by Fig. 8.5, in which the 'pure' P + D term is viewed as being followed by an appropriate low-pass filter.

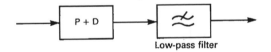

Fig. 8.5 Smoothed P + D action.

One of the simplest realizations of such a smoothed P + D device is illustrated conceptually in Fig. 8.6, where the low-pass filter is simply a first-order lag. The combined transfer function of such an element is therefore

$$\frac{K(1 + sT_d)}{1 + s\tau}$$

where τ is the time constant of the low-pass filter, which is to be chosen so as to

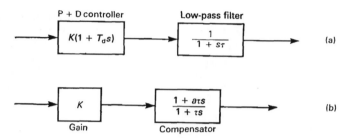

Fig. 8.6 Phase-lead compensation.

counteract the rising high-frequency response of Fig. 8.4 for frequencies $\omega > 1/\tau$

It is more convenient for design purposes to assume that the gain K is to be adjusted separately, and hence to model the smoothed P + D controller as shown in Fig. 8.6(b). Then the transfer function of the second block, with unity low frequency gain, can be written

$$\frac{(1 + a\tau s)}{(1 + \tau s)}$$

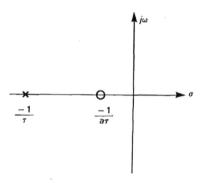

Fig. 8.7 Pole–zero plot of a phase-lead compensator.

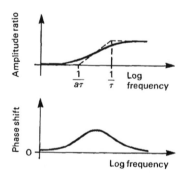

Fig. 8.8 Normalized frequency response of a lead compensator.

154

where the constant $a > 1$. The associated pole–zero plot is shown in Fig. 8.7, and the Bode plot in Fig. 8.8. As can be seen, the high-frequency amplitude ratio now approaches a constant value, rather than increasing indefinitely as in the pure P + D controller, while the phase, instead of approaching a constant +90° at high frequencies, peaks at an intermediate frequency and then returns towards zero. Because of this phase characteristic, a device with a frequency response like that of Fig. 8.8 is known as a *phase-lead compensator*.

Show using a labelled sketch how the frequency response of Fig. 8.8 can be derived from the combination of the Bode plots of the numerator and denominator of the transfer function.

Worked Example 8.2

Solution   See Fig. 8.9.

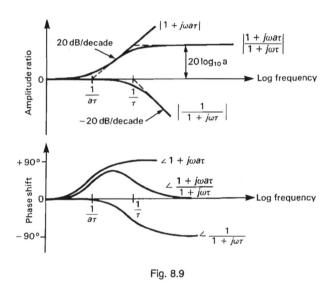

Fig. 8.9

In the years immediately following the Second World War, a great deal of effort was put into designing compensators which could be realized electrically using only resistors and capacitors.

One of the major advantages of the phase-lead compensator in the above form was that it could be realized as a simple, passive RC circuit. Nowadays, active phase compensators may be easily implemented electronically, using analogue operational amplifier circuits, or digital hardware or software. In such modern realizations the traditional lead compensator retains an attractive simplicity as a functional element for system design, even though the advantage of passive realization is no longer so important.

Approximate, or smoothed, P + D control may also be implemented in other ways. In particular, if digital hardware or software is used to provide derivative

Inductors were – and still are – comparatively bulky and difficult to manufacture to high standards, while active circuits at that time required expensive, and often unreliable, valves.

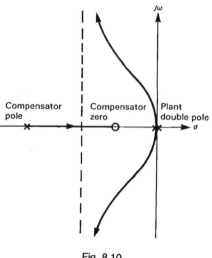

Fig. 8.10

Some elementary aspects of digital controllers will be considered below.

action, Fig. 8.5 can be realized using a digital low-pass filter which introduces the same sort of smoothing action as the first-order lag section of Fig. 8.6.

Returning to the s-plane model of the simple lead compensator, the important point to note is that the zero is closer to the $j\omega$ axis than the pole, as shown by the transfer function

$$\frac{1 + a\tau s}{1 + \tau s}$$

with $a > 1$.

If the compensator is properly designed, this zero can bring about the same sort of stabilizing and damping effects on a system as the zero introduced by a P + D controller, while the pole is sufficiently far away to have only a relatively minor influence. An illustrative root locus is shown in Fig. 8.10 for a double-integrator plant like the satellite of Chapter 6. Note, however, that if attempts are made to place the pole *too* far from the zero, the same noise problems begin to arise as with derivative action: as a is increased, and the distance between the compensator pole and zero becomes greater and greater, the lead compensator behaves more and more like a P + D controller. In frequency domain terms, the high-frequency attenuation becomes restricted to frequencies beyond those to which the plant can respond, and the compensator behaves very like a pure P + D controller.

For this reason a would not normally exceed 10 in any practical lead compensator.

As with much of control engineering, the precise choice of values of τ and a, and hence the pole and zero locations, is more a matter of experience and trial and error than of following strict design procedures. Certain general approaches have been found to be widely applicable, however.

One technique which is sometimes useful is to locate the compensator zero at the same position as an appropriate real plant pole (the compensator zero is then said to *cancel* the plant pole). The compensator pole is then positioned so as to give appropriate dynamic behaviour, perhaps with the aid of a study of the root locus.

Another, and perhaps more general, design method is to adopt the frequency response approach of Chapter 4. Then the designer attempts to choose values of a and τ so that the overall system has suitable gain and phase margins. The choice of such gain and phase margins will normally be made in the light of experience with similar systems.

In frequency response terms a phase lead compensator is used – as its name implies – to reduce the phase lag and hence improve the phase margin and the relative stability of an otherwise satisfactory design. As the Bode plot of Fig. 8.8 shows, however, the desired phase lead is accompanied by a change in gain over a range of frequencies. This has to be taken into account when calculating suitable compensator parameters: the trick is to calculate where the phase lead will be required once the open-loop frequency response incorporates the additional gain introduced by the compensator itself. (This will become clearer in the context of a design example below.)

In some cases, this extra gain can cause problems. The simplified Nichols plot of Fig. 8.11 illustrates the type of situation which can occur. The shape of the frequency response curve in the neighbourhood of the critical point ($-180°/0\,\mathrm{dB}$) means that a phase compensator giving a suitable phase margin results in a gain margin which is too low.

In some cases, therefore, lead compensation cannot satisfy both phase and gain margins.

The pole to be cancelled depends on the plant. A commonly-used guide is to choose the one with the longest time constant (closest to the origin) if the plant includes an integrator, or the second-longest time constant otherwise. The idea is then that the compensator pole and the remaining plant poles and zeros result in a root locus which will possess more desirable properties than that of the uncompensated system. It is difficult to generalize, however; the details of any cancellation approach will depend on the particular system. When using such a cancellation approach, it is important to remember that the dynamics of the plant itself are unchanged, even if poles appear to be cancelled from the complete transfer function. This can have important consequences for system behaviour, particularly on disturbance rejection, as mentioned briefly in Chapter 4 and taken up again later in this chapter.

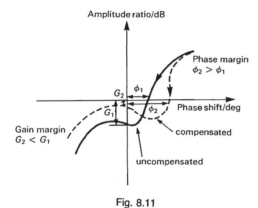

Fig. 8.11

Other types of phase compensation

If phase-lead compensation is viewed as approximate P + D control, then *phase-lag compensation*, corresponding to the transfer function

$$\frac{(1 + \tau s)}{(1 + a\tau s)}$$

where again $a > 1$, is an approximation to P + I control. The corresponding Bode plot and pole–zero diagram are shown in Fig. 8.12.

In this case, however, the name is a little misleading. The idea behind phase-lead compensation is indeed to incorporate phase lead into the system (or at least

Compare this Bode plot with the frequency response of a P + I controller given in Chapter 4, in the light of the following discussion.

to reduce phase *lag*), whereas the desirable feature of phase-lag compensation i not that it increases phase lag (which can only have an adverse effect on overa stability), but that it allows the loop gain to be increased without having too grea an effect on system dynamics. Suppose, for example, that the proportional gain c an uncompensated system is chosen to give a suitable transient response, but tha the value is too low for satisfactory steady-state performance. If a phase-lag con pensator is incorporated into the system, and the gain simultaneously increasec Fig. 8.12(a) shows that the effects of the increased gain will only be 'felt' at lo frequencies. For sufficiently high frequencies, the attenuation of the compensatc counteracts the increased proportional gain. So by appropriate choice of con pensator parameters, the transient behaviour (determined mainly by high fr quencies) can be maintained, while at the same time the steady-state errors ar reduced owing to the boost in low-frequency gain. In this sense, therefore, th lag compensator acts as an approximation to proportional + integral contro although, whereas an integrator can in theory *eliminate* steady-state error, a la compensator can only reduce it to a given level.

Finally, it is worth mentioning *lag–lead compensation* which, as its nam implies, is a combination of the two preceding forms of phase compensation. / lag–lead compensator may be viewed as an approximation to three-term contro

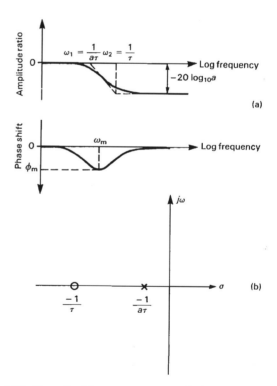

Fig. 8.12 Normalized frequency response of a lag compensator.

and again may be implemented in a number of ways, including passively as an RC network.

A compensator design example

The following practical design example concerns the frequency domain design of a lead compensator for a ship autopilot, although the approach can also be used for other types of compensator. The closed-loop system without the series compensator is modelled as shown in Fig. 8.13, where the actual heading is measured by a radiocompass. A steady-state analysis indicates that to satisfy error specifications under conditions of a steady turn (a tracking condition) a gain of $K = 5$ is required. With this gain, the open-loop frequency response is as shown on the Nichols chart of Fig. 8.14.

From the Nichols chart it can be seen that the system is marginally unstable for a gain which would satisfy steady-state specifications.

However, experience suggests that if a phase margin of 40–45° could be achieved, then dynamic behaviour would be satisfactory (providing the gain margin is not too low). Including a phase-lead series compensator in the forward path can have just this effect, as was made clear by the compensator's Bode plot of Fig. 8.8.

The next step is to choose appropriate values of a and τ. This is a little less straightforward than might at first appear, since, as mentioned above, the phase lead cannot be added independently of a change in gain. The increased gain introduced by the compensator will have the effect of raising the plot on the Nichols chart, while the phase lead will bend the curve round to the stable side of the $-180°/0\,\mathrm{dB}$ point. Sufficient phase lead is therefore required at a frequency which will itself depend on how much phase lead is added.

Although I will not derive them here, three useful design relationships may be deduced from the frequency response function of the lead compensator. Referring to Fig. 8.15 they are:

(1) $\omega_m = \sqrt{\omega_1 \omega_2}$

(where ω_m is the frequency corresponding to maximum phase angle ϕ_{max}, and $\omega_1 = 1/a\tau$ and $\omega_2 = 1/\tau$ are the two compensator corner frequencies);

(2) $\sin \phi_{max} = (a - 1)/(a + 1)$; and

(3) compensator amplitude ratio at

$$\omega_m = 10 \log \omega_2/\omega_1 \, \mathrm{dB}$$
$$= 10 \log a$$

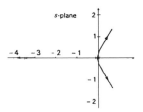

The root locus shows how quickly this system becomes unstable as K is increased.

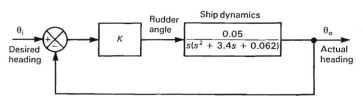

Fig. 8.13 A ship autopilot.

θ_i Desired heading

K

Rudder angle

Ship dynamics

$$\frac{0.05}{s(s^2 + 3.4s + 0.062)}$$

θ_o Actual heading

The Nichols plot can be constructed from the Bode plots of the individual terms in the factorized transfer function $0.05/s(s + 3.38)(s + 0.018)$, together with the gain $K = 5$. Note that the forward path possesses a pole at the origin, which ensures *zero* steady-state error to a step change in demanded input, and allows a *constant* steady-state error in response to a constant demanded rate of turn.

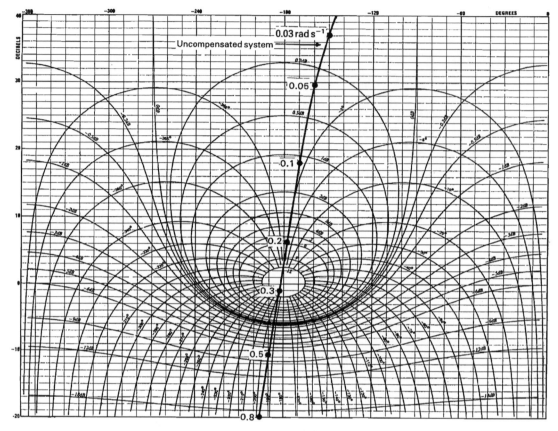

Fig. 8.14 Nichols plot of the uncompensated ship autopilot.

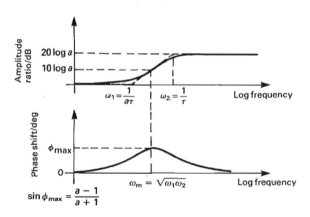

Fig. 8.15 Lead compensator design relationships.

Using these relationships appropriate values for the compensator parameters a and τ can be chosen. The Nichols plot suggests that a phase lead of 45–50° will be required at a frequency yet to be determined, so let us choose a compensator to have a ϕ_{max} of 50°.

Substituting $\phi_{max} = 50°$ into the second of the design relationships indicates a value of a of around 7 (since the entire procedure is approximate there is no need for greater precision). The extra gain factor at the frequency of maximum phase is therefore $10 \log 7 = 8.5$ dB. Returning to the Nichols plot, raising the entire curve by 8.5 dB means that the new phase margin will be measured to the point corresponding to about 0.45 rad s^{-1}. This is the frequency at which the 50° phase lead needs to be provided. The first of the design relationships can now be used to calculate ω_1 and ω_2. Specifically,

$$0.45 = \sqrt{7\omega_1^2}$$

Hence

$$\omega_1 = 0.17$$

and

$$\omega_2 = 7\omega_1 = 1.2$$

The required phase lead compensator will therefore have the frequency transfer function

$$\frac{(1 + j\omega/0.17)}{(1 + j\omega/1.2)}$$

Including this compensator in the forward path results in the final Nichols plot of Fig. 8.16. The phase margin is just over 40° and the gain margin approaching 20 dB. Testing and/or simulation would probably now be required to check performance, and if necessary slight modifications to parameters would need to be made.

As this example shows, compensator design of this type can involve quite intricate balancing of the various effects of the compensation. Some systems require several compensators, each with different frequency characteristics, in order to satisfy the specification. In such circumstances computer programs to plot frequency response curves quickly, and to simulate corresponding time domain behaviour, can save a great deal of time. Such tools are discussed in Chapter 12.

Disturbance rejection

So far in this chapter control system performance has been specified in terms of the closed-loop response to a step change in reference input – or a frequency domain equivalent such as amplitude and phase margins. Often, however, the dynamic response of the system to *disturbances* entering the loop is just as important as the response to changes in reference input. This is the case in many process control applications, where sudden changes to the reference input (set point) are uncommon, and the main task of the control system is to maintain the controlled variable(s) constant in spite of disturbances.

Figure 8.17 shows a system controlling a process which can be modelled as two,

The frequency is yet to be determined because the additional gain of the compensator will raise the curve on the Nichols plot. The phase margin of the final design will no longer be measured to the point corresponding to around 0.3 rad s^{-1}, but to a somewhat higher frequency on the curve.

Frequency response design like this can be carried out, with experience, using Bode plots alone: the details of gain and phase margins can be read from the separate amplitude ratio and phase plots, as shown in Chapter 4. However, the Nichols plot renders the effects on closed-loop response much clearer.

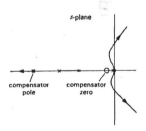

The behaviour of the lead compensator in this example may also be interpreted in root-locus terms. As in the cases discussed earlier, the compensator zero bends the dominant branches of the root locus to the left, stabilizing and speeding up the closed-loop response.

An example of how disturbance rejection dynamics were affected by the value of controller integral time in a control loop was given at the end of Chapter 4.

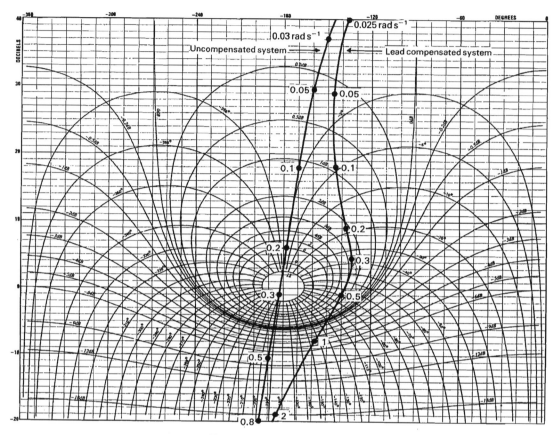

Fig. 8.16 Nichols plot of compensated autopilot.

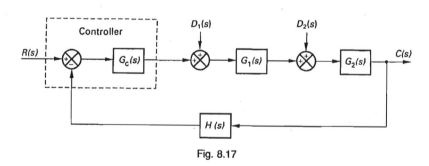

Fig. 8.17

cascaded, subsystems, G_1 and G_2. Disturbances d_1 and d_2 enter the loop at th
places shown. Applying the normal analysis procedures to this block diagram w
can write, after some manipulation:

$$C(s) = \frac{G_c(s)G_1(s)G_2(s)}{1 + G_c(s)G_1(s)G_2(s)H(s)}R(s) + \frac{G_1(s)G_2(s)}{1 + G_c(s)G_1(s)G_2(s)H(s)}D_1(s)$$

$$+ \frac{G_2(s)}{1 + G_c(s)G_1(s)G_2(s)H(s)}D_2(s)$$

This equation is much simpler to interpret than it looks! The first term on the right-hand side expresses the normal relationship between $R(s)$ and $C(s)$ in the absence of any disturbance entering the loop.

The other two expressions

$$\frac{G_1(s)G_2(s)}{1 + G_c(s)G_1(s)G_2(s)H(s)}$$

and

$$\frac{G_2(s)}{1 + G_c(s)G_1(s)G_2(s)H(s)}$$

can be interpreted as *disturbance transfer functions* modelling the additional effects of d_1 and d_2, respectively, on the system output c.

As was pointed out in Chapters 2 and 4, increasing the controller gain will tend to reduce the steady-state effect of both disturbances – providing the loop remains stable. (If the controller $G_c(s)$ includes integral action, the steady-state error from both sources of disturbance will, in theory at least, be completely eliminated.) Often, though, we need to know more than just the steady-state effect. Disturbance transfer functions are the key to understanding the *dynamic* effects of disturbances entering a control loop.

Consider the system of Fig. 8.18, representing an industrial water-heating system. Water is heated in a gas-fired boiler and passed to a mixing tank where it is mixed with an incoming stream of cold water. The temperature of the final mix is controlled by a proportional feedback controller which adjusts the gas flow to the boiler. Disturbance d_1 represents fluctuations in the gas pressure, while d_2 represents fluctuations in the temperature of the cold water supply to the mixing tank.

$$\frac{G_c(s)G_1(s)G_2(s)}{1 + G_c(s)G_1(s)G_2(s)H(s)}$$

is the closed-loop transfer function $C(s)/R(s)$ taking the usual form

$$\frac{\text{forward path transfer function}}{1 + \text{open-loop transfer function}}$$

Both disturbance transfer functions take the form

$$\frac{\text{'downstream' transfer function}}{1 + \text{open-loop transfer function}}$$

Note the implications of this form for designs in which a process pole is 'cancelled' from the closed-loop transfer function $C(s)/R(s)$ by a controller zero. Process poles remain in the disturbance transfer functions, since $G_c(s)$ does not appear as a multiplicative term in the numerator.

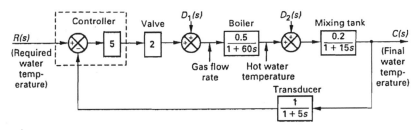

Fig. 8.18 Block diagram of a water-heating system.

163

Worked Example 8.3   What is the disturbance transfer function relating $C(s)$ to $D_2(s)$?

Solution   Using the above expression we can write the transfer function as

$$\frac{\dfrac{0.2}{(1 + 15s)}}{1 + \dfrac{1}{(1 + 60s)(1 + 15s)(1 + 5s)}}$$

or

$$\frac{0.2(1 + 60s)(1 + 5s)}{(1 + 60s)(1 + 15s)(1 + 5s) + 1}$$

Disturbance transfer functions can be plotted in frequency response form. Doing this by hand for a range of values of controller gain (or other parameter) would be very time-consuming. With computer tools of the type discussed in Chapter 12 however, it is easy to generate whole families of plots to investigate system performance. A family of Bode plots is illustrated in Fig. 8.19 for the water heating example. Amplitude ratios of the three frequency response functions of interest are shown for various values of controller gain K_{c1}. Figure 8.19(a) is the amplitude ratio of the closed-loop frequency response $C(j\omega)/R(j\omega)$, while part (b) and (c) give the corresponding plots for the two disturbance frequency response functions $C(j\omega)/D_1(j\omega)$ and $C(j\omega)/D_2(j\omega)$.

Consider first Fig. 8.19(a). As the controller gain is increased, the low-frequency amplitude response gets closer to 0 dB (unity) – that is, the steady-state error decreases. If the gain is increased sufficiently, however, the amplitude ratio exhibits a resonance peak, corresponding to a lower system damping ratio and increasingly oscillatory time domain behaviour.

This general feature of closed-loop behaviour has been encountered on numerous occasions before.

Next consider Fig. 8.19(b), which expresses the effect of d_1 on system output. Again, as the gain is increased, the steady-state effect of the disturbance decreases: the low-frequency amplitude ratio, expressing the steady-state sensitivity gets smaller and smaller. But as before, if the gain is increased sufficiently a resonance peak appears, indicating that the dynamic response of the system to a sudden *change* in disturbance is likely to be oscillatory. The disturbance transient will take longer to die away as the gain increases and the peak increases in magnitude. Figure 8.19(c), relating to disturbance d_2, shows similar characteristics but in an even more pronounced way. Moreover, resonance peaks appear in $|C(j\omega)/D_2(j\omega)|$ at values of gain for which there is no such peak in $|C(j\omega)/R(j\omega)|$ showing that a system may have an oscillatory response to d_2 *even if the response to changes in r is non-oscillatory.*

A resonance peak in a disturbance transfer function means that the system will be particularly sensitive to disturbances containing significant components at frequencies near the peak. Some control systems are designed specifically to reject disturbances with oscillatory components: examples include the position control of the antenna of a satellite ground station, which has to reject oscillatory wind loading; or systems designed to cancel out the effects of earthquakes on buildings. In such cases it is crucial to consider the disturbance rejection of the control system over an appropriate frequency range, not just the steady-state performance.

164

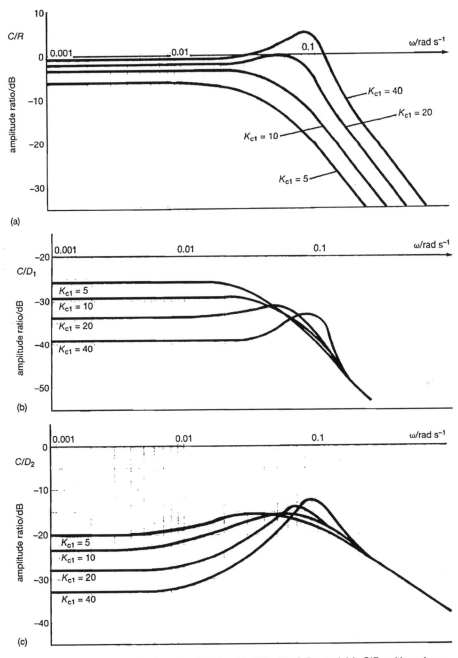

Fig. 8.19 Bode amplitude ratio plots for (a) C/R; (b) C/D_1; and (c) C/D_2 with various values of controller gain.

Cascade control

In a system like Fig. 8.18 it is often the case that a single controller cannot satisfy simultaneously all the steady-state, transient and disturbance rejection requirements. One strategy which can be used to improve performance is to use a second controller to form an inner loop.

Looking back at Fig. 8.18 it is clear that one of the problems is the boiler time constant of 60 s, which is long in comparison with the dynamics of the rest of the system. This implies a comparatively long delay before the effect of the disturbance d_2 on the output temperature c can be counteracted. By adding an inner loop around the boiler its time constant can be effectively reduced. What is more, this inner loop will tend to reject disturbance d_1, as well as speeding up the outer loop's reaction to d_2.

The strategy is known as cascade control, and is illustrated in Fig. 8.20. The gains of the two controllers in Fig. 8.20 have been chosen so that the boiler loop time constant is 10 s and the overall low-frequency loop gain is 5. Figure 8.21 contrasts the frequency domain performance of the cascade scheme (bold curve) with that of the original system using the same loop gain. As can be seen, introducing the second controller has (i) increased the overall system bandwidth; (ii) greatly improved d_1 rejection characteristics (by 15 dB at low frequencies); and (iii) reduced the resonance peak in the d_2 response by several dB.

It can be shown – somewhat counter-intuitively – that the greatest performance improvement with cascade control is obtained by putting an inner loop around the component with the *shortest* time constant. This is not possible in the water-heating example, as can be seen by looking at the intermediate variables in Fig. 8.18.

Cascade schemes are widely used in process control applications. For such a scheme to be possible, of course, there has to be an appropriate intermediate variable to measure and control. Cascade control often involves an inner loop at the 'supply side' of a process, controlling the position of a valve (as in the example here) or the flow from a pump.

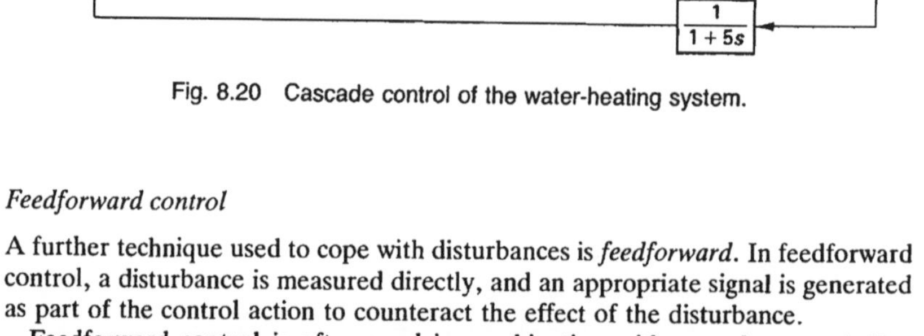

Fig. 8.20   Cascade control of the water-heating system.

Feedforward control

Feedforward control was briefly mentioned in Chapter 1.

A further technique used to cope with disturbances is *feedforward*. In feedforward control, a disturbance is measured directly, and an appropriate signal is generated as part of the control action to counteract the effect of the disturbance.

Feedforward control is often used in combination with cascade control. Returning to Fig. 8.21(c), it can be seen that, although cascade control reduced the resonance peak in the system response to d_2, it did not improve the steady-state

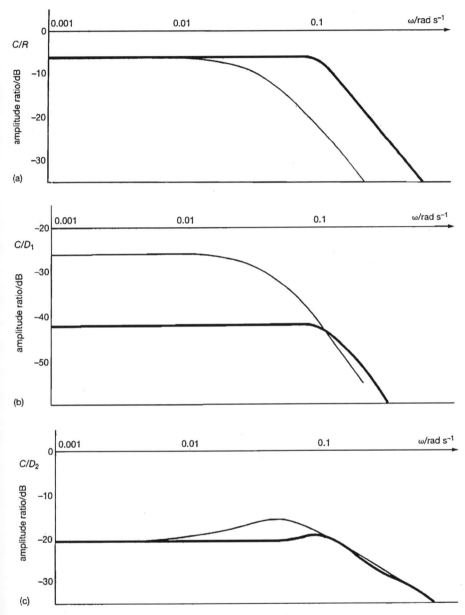

Fig. 8.21 Bode amplitude ratio plots for the cascade control system (bold curves) showing improved system dynamics and disturbance rejection properties.

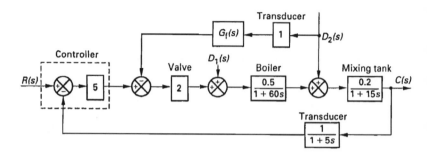

Fig. 8.22   Feedforward control of the water-heating system.

disturbance rejection – the low-frequency gains are the same with and withou cascade control. Figure 8.22 shows the use of feedforward to counteract d (without a cascade controller). The disturbance (in this case a change to the temperature of the cold water entering the mixing tank) is measured, and the value is used to adjust the signal sent to the gas valve positioner. The processing carried out on the disturbance signal is represented by the feedforward transfer function $G_F(s)$. Although *perfect* feedforward compensation for the effect of disturbance is never possible, significant improvement in disturbance rejection can be achieved – always providing reasonably good models of the process are available.

Worked Example 8.4

If you find this argument unconvincing, label the signals at appropriate points in Fig. 8.22 and carry out the necessary algebraic manipulations of the transforms to show that $G_F(s) = 1/G_1(s)$, where $G_1(s)$ represents the valve and boiler combined. For a more extensive treatment of disturbance rejection see *Cascade and Feedforward Control* (Unit 16 of the Open University Course *T394 Control Engineering*) on which this section is based.

What feedforward transfer function $G_F(s)$ would be required in Fig. 8.22 to give the best possible compensation for the effects of d_2? What sort of device could be used to realize $G_F(s)$, assuming that the dynamics of the transducer measuring d can be ignored?

Solution $G_F(s)$ has to 'undo' the dynamic effects of the valve and boiler combined. Since the transfer function of the valve/boiler combination is $1/(1 + 60s)$ the ideal $G_F(s)$ would be simply $(1 + 60s)$. This could be realized by a P + I controller with proportional gain 1 and derivative time 60 s.

Summary

This chapter has introduced the three-term or PID controller, which combines the desirable features of the P + I and P + D controllers described previously. PID controller settings can be determined using the design techniques of previous chapters, or, for plants which behave in a particular well-defined way, by means of empirical rules.

The principles of phase compensation were also described. A frequency domain design procedure was outlined for a phase-lead compensator; the same general principles apply to the design of other types of compensator and controller.

Finally, some aspects of disturbance rejection were considered, with a brief discussion of the characteristics of cascade and feedforward control.

Problems

8.1 The following table gives the results of an open-loop step response test. Suggest values of PID controller parameters to obtain a quarter-amplitude closed-loop response. (The steady-state response is 10 units.)

Time/s	Response (normalized to unit step input)
0	0
1	0.3
2	0.8
3	1.9
4	3.1
5	4.1
6	5.1
7	6.0
8	6.9
9	7.5
10	8.0
12	8.5
15	8.9
20	9.2
25	9.4

8.2 Using a frequency domain approach, investigate the effect of adding derivative action to the temperature control system design example of Chapter 4. Set the derivative time T_d equal to $T_i/4$. What are the effects of adding derivative action on: (a) the closed-loop frequency response; (b) the closed-loop step response of the system?

8.3 For the system of Fig. 8.23 choose values of gain and compensator parameters such that:
   (a) the peak overshoot of the step response is restricted to 15%;
   (b) the steady-state position error in response to an input ramp of $1 \, \text{rad s}^{-1}$ is not more than 0.025 rad;
   (c) the phase margin is around 55°.

8.4 Derive the phase-lead compensator design relationships given on page 159. (*Hint*: use symmetry arguments based on the Bode plot, and then standard trigonometrical relationships.)

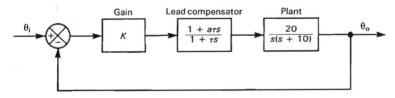

Fig. 8.23

8.5 The Ziegler and Nichols closed-loop method is used to determine the settings of a controller. The ultimate value of gain K_U is 30, and the ultimate period of oscillation T_U is 32 s. What would be the recommended settings for (a) a proportional controller, (b) a P + I controller, and (c) a P + I + D controller?

Digital control I. Discrete system models

9

□ To explain the significance of sampling, quantization and data interpolation in computer-based control systems.
□ To introduce the z-transform as a model of discrete signals.
□ To introduce the difference equation as a model of discrete linear systems.
□ To introduce discrete integrators, differentiators and filters.
□ To extend the concept of the transfer function to discrete linear systems, and to present z-plane pole–zero models.

Objectives

Introduction

It has been noted already in this book that digital electronics and digital computers play a vital role in modern control engineering. Sophisticated microprocessor-based control systems are incorporated into machine tools, industrial robots, and domestic audio equipment. Modern process controllers are also overwhelmingly digital in nature, and offer the user many features in addition to the classic three-term control algorithm discussed in the previous chapter. Furthermore, the widespread adoption of information technology in industrial process plant and manufacturing systems has enabled individual digital control loops to be integrated into complex, computer-based management systems.

Details of such large-scale control systems are beyond the scope of this book. This chapter and the next concentrate on developing a new set of modelling tools used in the analysis and design of individual digital feedback loops. The new techniques are very closely related to the Laplace transform and the s-plane pole–zero models introduced in Chapter 5.

Information technology is often defined as the convergence of telecommunications and computing. In process control and manufacturing, this trend has meant that digital communication techniques have been combined with computer-based processing and monitoring, resulting in extremely powerful and flexible systems.

Sampling and digitization

The theoretical development of the preceding chapters rests on the tacit assumption that all signals in the systems under consideration can vary continuously with time. So, for example, the output signal from a transducer (most often a voltage or current) has been assumed to correspond at all times to the variable being measured, subject only to the transducer's dynamic response and accuracy. Such signals are known as analogue signals: in a temperature-to-voltage transducer, for example, the output voltage is a continuous, electrical analogue of the temperature being measured. Things are rather different, however, once digital processing is used in a control system.

Consider Fig. 9.1, which represents a general, single-loop, digital control

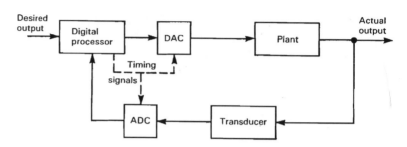

Fig. 9.1 A digital control system.

It is becoming increasingly common for transducers to produce an output already coded in digital form, in which case a separate ADC is not necessary in Fig. 9.1.

scheme. Here the measured value of the controlled variable is converted a regular intervals into a digital representation by means of an analogue-to-digita converter (ADC).

These digitized *samples* of the controlled variable are made available to th digital processor, which uses them, together with supplied information about th desired output, to compute the necessary control action. The processor may b implemented either in hardware, or as a computer program. The output of th processor may be viewed as a sequence of numerical values which, after conver sion to analogue form by means of a digital-to-analogue converter (DAC) provides the control action to drive the plant.

The major difference between the analogue systems considered so far, and th digital control system of Fig. 9.1, is that the latter involves the following thre distinct, although related, effects.

(1) *Sampling*. When a continuous signal is sampled, its value is measured a particular instants of time only. Samples are usually equally spaced in time. If th sampling interval is T seconds, then the number of samples per second, generall referred to as the sampling frequency f_s, is $1/T$, as illustrated in Fig. 9.2(a). Th behaviour of the signal between sampling instants is not detected. Because th sampled signal is defined at specific instants only, it is known as a *discrete-tim* signal. Note, however, that the samples of the figure are still (in theory) precis measures of the analogue waveform at the sampling instants.

The sampled signal of Fig. 9.2(a) is a discrete signal, but not a digital one, because it is *not* restricted to one of a finite set of values. The distinction between digital and discrete signals is often glossed over in the technical literature. Most of the theory of the next three chapters is concerned with discrete models: the effects of *digitization* have to be modelled separately if necessary.

(2) *Quantization*. Once a signal has been sampled, its continuously variable valu must be coded using a finite number of bits. The number of bits available to cod a sample represents a fundamental limitation on resolution: for example, if eacl sample is coded using 4 bits, then a maximum of $2^4 = 16$ different signal levels ar available to cover the range of the signal. Signal values lying between thes quantization levels must be coded by rounding to the nearest level (Fig. 9.2(b)) introducing a certain unavoidable error in the representation of the signal. Th quantized sample values are limited to a finite set of numerical values and thu constitute a *digital* signal. Sampling and quantization together are often referre to as digitization.

(3) *Data interpolation*. The required control action calculated by the digita processor as a sequence of digital values must ultimately be converted into ar electrical or other physical signal in order to drive a valve, motor or othe

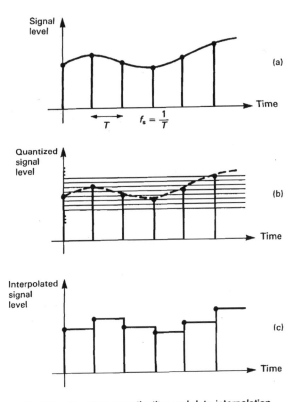

Fig. 9.2   Sampling, quantization and data interpolation.

actuator. This is normally carried out using some form of interpolation between the sampling instants – although discrete pulses may be used, as in the case of a stepper motor. Figure 9.2(c) illustrates the simplest, and commonest, form of *interpolation*, in which the value of a sample is held constant between sampling instants. Most DACs produce an output which approximates to such a 'staircase' waveform.

The *resolution* of a DAC is the smallest change in its output, usually expressed as a percentage of full scale. For an n-bit converter, therefore:

resolution $= 100/2^n \%$

Resolution is often specified as dynamic range:

dynamic range in dB $= 20\log_{10}2^n \approx 6n$

The DAC output, although digital, is no longer a discrete-time signal, since it has a well-defined value *between* sampling instants.

What are the resolution and dynamic range of a 12-bit DAC?

Worked Example 9.1

Solution   Substituting in the above expressions gives a resolution of 0.024% and a dynamic range of 72 dB.

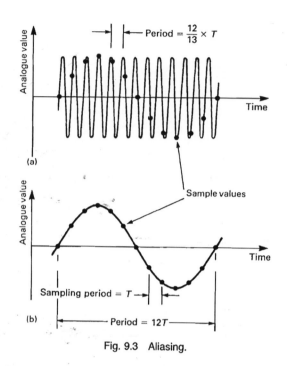

Fig. 9.3 Aliasing.

The processes of sampling, quantization and data interpolarization have a profound effect on the behaviour of a digital control system. Consider first the sampling process illustrated in Fig. 9.3(a). In this example the sampling frequency is close to the frequency of the sinewave being sampled, and an ambiguity arises: the samples could just as easily represent the completely different sinewave of Fig. 9.3(b). This phenomenon is known as aliasing, and the lower-frequency sinusoid is said to be the alias of the higher-frequency one. Clearly, there is a lower limit to the sampling rate if all frequencies present in a continuous signal are to be properly represented in the sampled version. Sampling a controlled variable at too low a rate results in a loss of information about the way the variable changes with time, and without this information effective control is impossible.

The general problem of the transmission of information in sampled form has been studied in detail, and the *sampling theorem* expresses the theoretical lower limit to the sampling rate if aliasing is to be avoided. One statement of the theorem is as follows:

> If the frequency components present in a continuous signal extend from 0 to B Hz, then the signal can be completely represented by, and reconstructed from, a sequence of equally spaced samples, provided that the sampling frequency exceeds $2B$ samples per second.

In other words, the highest frequency component of a signal must be sampled more often than twice per cycle.

174

One important practical consequence of aliasing is that high-frequency noise present in a transducer output signal may be misinterpreted as much lower-frequency variations in the measurand. To the digital processor of Fig. 9.1, such aliases would be indistinguishable from real, low-frequency changes in the controlled variable, and a corresponding and unwanted control action would be generated. For this reason, analogue-to-digital converters are usually preceded by a low-pass *anti-aliasing filter* to remove high-frequency components of the transducer output.

The effects of digitization are also important. The choice of converter word-length in a digital system is determined by various practical constraints. For example, it is clearly pointless to choose converters with a resolution much greater than that of the transducer measuring the controlled variable. The cost of data converters also increases rapidly with the number of bits used per sample. On the other hand, if quantization is too coarse, additional *quantization noise* is introduced, which can degrade performance.

In the majority of systems, data interpolation at the controller output produces a staircase waveform of the type shown in Fig. 9.2(c). If the quantization interval of the DAC is small enough, and the sampling rate high enough (compared with overall system dynamics), then the stepped nature of such waveform can be neglected. For lower sampling rates, however, the phenomenon has to be taken into account. But what, exactly, is a 'high enough' sampling rate?

In a control system, it is difficult to specify the 'highest' frequency component in a signal. So although the sampling theorem states that the sampling rate must be at least twice this frequency, it is not obvious how the theoretical result can be translated into useful criteria for choosing the sampling rate in a control system. In practice, various rules of thumb have been devised. Since the system closed-loop bandwidth is a measure of the range of frequencies which will be present in a control system, it is often used in such guidelines. At sufficiently high sampling rates (of the order of 40–50 times the closed-loop 3 dB bandwidth, say), the parameters of digital controllers can often be chosen and adjusted exactly as if the system were continuous, without any noticeable effect on system behaviour. This is not the only approach, however. Indeed, very high sampling rates may not always be feasible – or may not be the best engineering solution. For example, high rates may require expensive hardware to carry out all the necessary computation in the space of one sampling interval. Excessively high sampling rates can also cause problems with the control algorithm itself. (For example, if a variable is sampled so frequently that successive samples differ only slightly, any algorithm involving taking the *difference* between them may be subject to considerable numerical error.) In fact, it is quite possible to design digital control systems with sampling intervals which are long enough to have a significant effect on system behaviour, while at the same time being short enough to avoid aliasing problems. Figures quoted in the literature suggest that in suitably favourable circumstances control system sampling rates can be as low as 5 to 10 times the closed-loop 3 dB bandwidth.

To model and design systems using sampling rates towards the lower end of the practical range requires a new set of tools for taking into account the sampling and interpolation processes. The remainder of this chapter will introduce models of discrete signals and the systems which process them; Chapters 10 and 11 will

High-frequency noise is often present in analogue control systems too. Here, however, unwanted control action will at least be generated at the same comparatively high frequencies as the noise itself, to which the plant will be relatively insensitive. In a digital system, on the other hand, aliased noise may well fall into the range of frequencies to which the plant can respond and is therefore potentially more serious.

An alternative rule of thumb is that there may be as few as 2–4 samples during the rise time of the step response.

175

then apply these new tools to closed-loop digital control systems like that of Fig. 9.1.

The z-transform

The theory of z-transforms was developed in the 1940s and 1950s to provide analytical techniques for sampled data systems similar to the Laplace transform methods available for continuous systems. The z-transform provides an algebraic way of representing a sequence of numbers and describing the relationship between different sequences in a sampled data system.

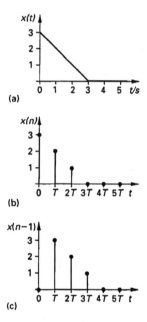

Fig. 9.4 Sampling an analogue signal: (c) shows the sequence of (b) delayed by one sampling interval.

If the signal $x(t)$ shown in Fig. 9.4(a) is sampled at one-second intervals, starting at $t = 0$, the sample sequence $x[n]$ shown in Fig. 9.4(b) is obtained:

$$x[n] = 3, 2, 1, 0, 0, \ldots$$

The z-transform of this sequence is represented by $X(z)$ and is defined by the sum

$$X(z) = 3z^0 + 2z^{-1} + 1z^{-2} + 0z^{-3} + 0z^{-4} + \ldots$$

Formally, the z-transform of the sequence $x[n] = x_0, x_1, x_2, x_3, \ldots$ is given by

$$X(z) = \sum_{i=0}^{\infty} x_i z^{-i}$$

In other words, the power of z^{-1} 'tags' the position of the sample in the sequence. For conciseness, this z-transform would normally be written simply

$$X(z) = 3 + 2z^{-1} + z^{-2}$$

The z notation is particularly useful for handling delays. For example, the original sequence delayed by one sampling period, as in Fig. 9.4(c), can be written:

$$0, 3, 2, 1, 0, 0, \ldots$$

and is modelled by the z-transform

$$0 + 3z^{-1} + 2z^{-2} + z^{-3}$$

or simply

$$3z^{-1} + 2z^{-2} + z^{-1}$$

This is just the z-transform of the original sequence multiplied by z^{-1}. Multiplication by z^{-1} denotes a delay of one sampling period, and so z^{-1} is sometimes known as the *delay operator*. The delay property of z^{-1} can be expressed formally as follows:

The apparently illogical use of z^{-1}, rather than z, to indicate a delay of one sampling period has its roots in the historical development of the subject.

> If a sequence $x[n]$ is represented by the transform $X(z)$ then the sequence delayed by k sampling periods, $x[n - k]$, has the transform $z^{-k}X(z)$.

(a) Write down the z-transforms of the sequences shown in Fig. 9.5.
(b) Write down the sequences corresponding to the following z-transforms:
   (i) $z^{-1} - 1.5z^{-2} + 3z^{-4}$
   (ii) $z^{-2} - z^{-4} + z^{-6}$

Worked Example 9.2

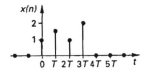

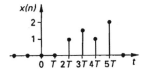

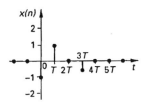

Fig. 9.5

177

Solution

(a) (i) The sample sequence is

$$1, 1.5, 1, 2, 0, 0, 0, \ldots$$

Writing down an appropriately weighted power of z^{-1} for each term, gives the z-transform

$$1 + 1.5z^{-1} + z^{-2} + 2z^{-3}$$

(ii) This sample sequence is the sequence of part (i) delayed by two sampling intervals. Hence its transform is found by multiplying that of sequence (i) by z^{-2}, giving

$$z^{-2}(1 + 1.5z^{-1} + z^{-2} + 2z^{-3})$$

or

$$z^{-2} + 1.5z^{-3} + z^{-4} + 2z^{-5}$$

(iii) The sequence is

$$-1, +1, 0, -0.5, 0, 0, 0 \ldots$$

giving the transform

$$-1 + z^{-1} - 0.5z^{-3}$$

(Note that it is not necessary to include a term $0z^{-2}$ representing the sample of value zero at $t = 2T$.)

(b) (i) The transform

$$z^{-1} - 1.5z^{-2} + 3z^{-4}$$

has no term representing a sample at $t = 0$. The first sample in the sequence is therefore 0. Similarly, the sample at time $t = 3T$ is also 0. The entire sequence is:

$$0, 1, -1.5, 0, 3$$

(ii) In this case the samples at $t = 0$, T, $3T$ and $5T$ are all 0. The sequence is therefore

$$0, 0, 1, 0, -1, 0, 1$$

Z-transform models of signals can also be found if we know the mathematical form of the sampled signal. Consider, for example, a signal modelled as a decaying exponential function $x(t) = e^{-t}$. What is the z-transform of the sampled version of this signal?

Figure 9.6 shows the signal sampled every T seconds. The value of $x(t)$ at time $t = 0$ is 1, so $x_0 = 1$. At the second sampling instant $t = T$ the value of $x(t)$ has fallen to e^{-T}, so $x_1 = e^{-T}$. Repeating this process indefinitely at intervals of T gives the sequence of samples

$$x[n] = 1, e^{-T}, e^{-2T}, e^{-3T}, \ldots$$

which can immediately be transformed to

$$X(z) = 1 + e^{-T}z^{-1} + e^{-2T}z^{-2} + e^{-3T}z^{-3} + \ldots$$

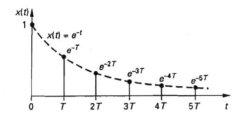

Fig. 9.6   A sampled, decaying exponential signal.

$X(z)$ is an infinite geometric series – that is, a series of the form

$$S = 1 + r + r^2 + r^3 + \ldots$$

where, in this case, $r = e^{-T}z^{-1}$.

Textbooks that deal with the detailed mathematics of digital control theory are often much concerned with the convergence properties of such series. I shall not go into this aspect here. It is sufficient for our purposes that under certain conditions, which apply to all the transforms in this text, the sum S of the geometric series can be shown to be

$$S = \frac{1}{1 - r}$$

Using this result the z-transform of the sampled decaying exponential signal can be expressed as

$$X(z) = \frac{1}{1 - e^{-T}z^{-1}}$$

Multiplying numerator and denominator by z gives the more usual form

$$X(z) = \frac{z}{z - e^{-T}}$$

Note that this transform of the sampled exponential includes the sampling period T explicitly. This reflects the fact that the value of T will affect the values of the individual samples in the original sampled data sequence.

(a) Write down the z-transform of the decaying exponential $x(t) = e^{-t}$ sampled at a frequency of 10 Hz.

(b) A unit step function is sampled every T seconds. Work out the z-transform of the resulting sample sequence, assuming that the value of the unit step at time $t = 0$ is 1. What would be the z-transform of a sampled step delayed by T seconds?

Worked Example 9.3

Solution

(a) The z-transform of the signal $x(t) = e^{-t}$ sampled at intervals of T seconds is

$$X(z) = \frac{z}{z - e^{-T}}$$

Substituting 0.1 for T, we have

$$X(z) = \frac{z}{z - e^{-0.1}} = \frac{z}{z - 0.9}$$

(b) A sampled unit step function consists of the sequence

$$1, 1, 1, 1, 1, \ldots$$

with the z-transform

$$X(z) = 1 + z^{-1} + z^{-2} + z^{-3} + \ldots$$

We can use the formula for the sum of a geometric series

$$S = \frac{1}{1 - r}$$

In this case $r = z^{-1}$, and

$$X(z) = \frac{1}{1 - z^{-1}}$$

Multiplying numerator and denominator by z gives

$$X(z) = \frac{z}{z - 1}$$

If this sequence is delayed by one sampling interval, then its transform is multiplied by z^{-1}. The transform of the delayed sampled step is therefore

$$z^{-1}X(z) = \frac{z^{-1}z}{z - 1} = \frac{1}{z - 1}$$

Fortunately, z-transforms do not have to be evaluated from first principles each time, although it is important to understand how they are derived. Transforms of

Table 9.1

Continuous signal $x(t)$	Laplace transform $X(s)$ of continuous signal	z-transform $X(z)$ of sampled signal
$u(t)$ [step]	$\dfrac{1}{s}$	$\dfrac{z}{z - 1}$
t [ramp]	$\dfrac{1}{s^2}$	$\dfrac{Tz}{(z - 1)^2}$
e^{-at}	$\dfrac{1}{s + a}$	$\dfrac{z}{z - e^{-aT}}$
$1 - e^{-at}$	$\dfrac{a}{s(s + a)}$	$\dfrac{z(1 - e^{-aT})}{(z - 1)(z - e^{-aT})}$
$\sin \beta t$	$\dfrac{\beta}{s^2 + \beta^2}$	$\dfrac{z \sin \beta T}{z^2 - (2 \cos \beta T)z + 1}$
$e^{-at} \sin \beta t$	$\dfrac{\beta}{(s + a)^2 + \beta^2}$	$\dfrac{ze^{-aT} \sin \beta T}{z^2 - (2e^{-aT} \cos \beta T)z + e^{-2aT}}$

common sampled functions are tabulated in the same way as Laplace transforms. Table 9.1 lists some common signal models, their Laplace transforms and the z-transforms of the sampled signals.

Using Table 9.1 calculate the z-transforms of the following sampled signals:

Worked Example 9.4

(a) A ramp of slope 2 sampled every second.
(b) The sampled step response of the system whose transfer function is $1/(1 + 2s)$, sampled at 3 Hz.

Solution
(a) From Table 9.1 we see that a sampled unit ramp (with slope 1) has a z-transform $Tz/(z - 1)^2$.

For a ramp of slope 2, $x(t) = 2t$. The z-transform of a sampled version of this signal is therefore

$$X(z) = \frac{2Tz}{(z - 1)^2}$$

Substituting $T = 1$ s gives

$$X(z) = \frac{2z}{(z - 1)^2}$$

(b) The transform of the step response of a system is equal to the system transfer function multiplied by $1/s$, the transform of the step. In this case, therefore, the step response has the transform

$$X(s) = \frac{1}{s(1 + 2s)}$$

Table 9.1 has an entry for $a/[s(s + a)]$, so we need to manipulate $X(s)$ into this form. Dividing numerator and denominator of the equation for $X(s)$ by 2 gives

$$X(s) = \frac{0.5}{s(s + 0.5)}$$

which is of the desired form, where $a = 0.5$. The corresponding z-transform is therefore

$$X(z) = \frac{z(1 - e^{-0.5T})}{(z - 1)(z - e^{-0.5T})}$$

Substituting $T = 0.33$ s gives

$$X(z) = \frac{z(1 - 0.846)}{(z - 1)(z - 0.846)}$$

or

$$X(z) = \frac{0.15z}{(z - 1)(z - 0.846)}$$

(Note that converting $1/s$ and $1/(1 + 2s)$ separately into z-transforms and then multiplying them together will give an incorrect result.)

Alternatively, note immediately that the step response of the system with the transfer function

$$\frac{1}{1 + 2s}$$

is $(1 - e^{-t/2})$. The table has the entry $x(t) = 1 - e^{-at}$, so putting $a = 0.5$ in the corresponding entry for $X(z)$ gives the same result.

Worked Example 9.5 Using Table 9.1, write down the first few samples of the sequences corresponding to the transforms:

(a) $\dfrac{3z}{(z-1)}$

(b) $\dfrac{z}{2(z-1)^2}$

Sketch the sample sequence corresponding to (b).

Solution

(a) From the table $z/(z-1)$ is the transform of a sampled unit step. So the expression $3z/(z-1)$ is the transform of a sampled step of height 3, and the sequence is 3, 3, 3, ...

(b) From Table 9.1

$$\frac{Tz}{(z-1)^2}$$

is the transform of a sampled unit ramp $x(t) = t$. Hence the expression

$$\frac{z}{2(z-1)^2} = \frac{0.5z}{(z-1)^2}$$

can be thought of as the transform of a unit ramp sampled every 0.5 seconds and the sequence is

$$0, 0.5, 1, 1.5, 2, 2.5, \ldots$$

This sequence is sketched in Fig. 9.7.

A specific z-transform corresponds to a sequence of numbers. Only with a knowledge of the sampling interval can a time scale be added. The transform $0.5z/(z-1)^2$ can also be thought of as corresponding to a ramp of slope 0.5, sampled every second, instead of a ramp of slope 1, sampled every 0.5 s. The same sequence results – and in neither case can we say anything about real-time behaviour.

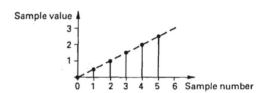

Fig. 9.7 Samples corresponding to the transform $z/2(z-1)^2$.

Transfer function models

Just as the frequency response function $G(j\omega)$ and the Laplace transfer function $G(s)$ are used to represent the relationship between input and output signals in a continuous linear system, so a transfer function in z can be used to represent the relationship between input and output *sequences* in a discrete linear system. By a discrete linear system is meant one which obeys the principle of superposition. That is:

If an input sequence $x_1[n]$ produces the output sequence $y_1[n]$, and an input $x_2[n]$ produces the output $y_2[n]$, then the input $x_1[n] + x_2[n]$ produces the output $y_1[n] + y_2[n]$.

Recall from Chapter 5 that, assuming the initial quiescence condition, a transfer function is defined as:

$$\text{transfer function} = \frac{\text{transform of output signal}}{\text{transform of input signal}}$$

You may recognize superposition as one of the properties of the Laplace transform, described in Chapter 5. In fact, the principle is a general definition of linearity, which applies to both continuous and discrete systems. For a more detailed discussion of linearity see Meade and Dillon (1991).

In a sampled data system, therefore, a transfer function in z expresses the way the system modifies the input sequence to produce the output sequence. If $R(z)$ and $C(z)$ represent the z-transforms of the input and output sequences of a system, respectively, as in Fig. 9.8(a), then the transfer function of the system is

$$G(z) = \frac{C(z)}{R(z)}$$

As with Laplace transforms the overall transfer function of a series, or cascade, connection of system blocks is given by the product of the individual transfer functions. In Fig. 9.8(b), therefore, the overall transfer function is

$$\frac{Y(z)}{R(z)} = \frac{C(z)}{R(z)} \times \frac{Y(z)}{C(z)}$$
$$= G(z)H(z)$$

If we know the transfer function of a system then we can calculate the output samples produced by the system in response to a given input sequence. The input–output relationship can be written

$$\text{transform of output} = \text{transfer function} \times \text{transform of input}$$

that is

$$C(z) = G(z) \times R(z)$$

Exploiting this relationship, standard tables are sometimes used to calculate the response sequences to the more common test inputs, such as the sampled sine-wave or sampled unit step. More often, however, computer-based tools are used to generate such response sequences directly.

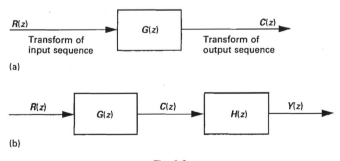

Fig. 9.8

183

The unit sample response sequence

Suppose that the input to a digital system is the simple sequence

$$r[n] = 1, 0, 0, 0, 0, \ldots$$

This is called the unit sample sequence. The z-transform of the sequence is jus $R(z) = 1$. If this sequence is applied to the input of a system with a transfe function $G(z)$ then the transform of the resulting output is given by

$$C(z) = G(z) \times 1 = G(z)$$

In other words the z-transform of the response of a sampled data system to a un sample input sequence is equal to the transfer function of the system.

A knowledge of the unit sample response sequence can therefore be used t derive a system transfer function.

Worked Example 9.6 What are the transfer functions in terms of z of the digital processors with th following unit sample response sequences?

(a) $0, 2, -1, 0, 0, 0, \ldots$
(b) $1, \frac{1}{3}, \frac{1}{9}, \frac{1}{27}, \ldots$

Solution
(a) The transform of the unit sample response, $G(z)$, is identical to the transfe function of the digital system. Hence,

$$G(z) = 2z^{-1} - z^{-2}$$

To express this in terms of powers of z, rather than z^{-1}, we multiply throug] by z^2, giving

$$G(z) = \frac{2z - 1}{z^2}$$

(b) In this case

$$G(z) = 1 + \tfrac{1}{3}z^{-1} + \tfrac{1}{9}z^{-2} + \tfrac{1}{27}z^{-3} + \ldots$$

First we sum this geometric series as

$$G(z) = \frac{1}{1 - \frac{1}{3}z^{-1}}$$

Multiplying throughout by z gives

$$G(z) = \frac{z}{z - \frac{1}{3}} \approx \frac{z}{z - 0.33}$$

Difference equation models

The relationship between the transfer function and the unit sample respons sequence is a very useful one, to which we shall return later. An alternative rout to the transfer function in z of a discrete linear system, however, is what is known as the *difference equation* model of the system. Difference equations are to discrete systems what differential equations are to continuous systems; both types of equation can easily be converted to a corresponding transfer function model. Let us begin by looking at difference equation models of discrete approximations to differentiation and integration, operations which we know can be exploited to advantage in control systems.

Remember that in the control context, integration and differentiation are not used because of their precise mathematical characteristics, but to provide desired system effects – derivative action to improve dynamic behaviour, for example, or integral action to eliminate steady-state error. How close a discrete approximation is to the *ideal* mathematical operation is, for control engineers, a secondary consideration – in contrast to the use of discrete algorithms for the numerical solution of differential equations. The discrete approximations used in control are designed to produce the required system effects without excessive computation.

Discrete differentiation

Suppose that an error signal is available as a sequence of samples $e[n]$, as shown in Fig. 9.9, where e_n represents the nth sample, taken at time nT. Then an approximation d_n to the value of the derivative de/dt at time nT can be found simply by calculating the gradient of the line joining two consecutive samples. The operation can be represented by the following *difference equation*

$$d_n = \frac{(e_n - e_{n-1})}{T}$$

The operation is represented schematically in Fig. 9.10, where the block labelled T represents a delay of T seconds – that is, the relevant sample value is stored for the duration of one sample interval and then released for further processing. The functions of the multiplier and summation elements should be clear from the context. If the input error sequence is $e[n]$, then the output of the delay block is $e[n-1]$, the same sequence delayed by one sampling interval. The output sequence $d[n]$ is, at any given sampling instant, an approximation to the derivative of the error signal. We can write, in terms of the sequences as a whole,

$$d[n] = \frac{(e[n] - e[n-1])}{T}$$

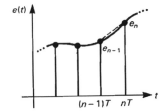

Fig. 9.9   Using samples to approximate the derivative.

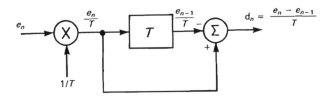

Fig. 9.10   A functional block diagram of discrete differentiation.

185

The z-transform models developed in the previous sections now come into their own. Writing the transform of the error sequence as $E(z)$, that of the derivative sequence as $D(z)$, and that of the delayed error sequence as $z^{-1}E(z)$ we have

$$D(z) = \frac{E(z) - z^{-1}E(z)}{T}$$

or

$$D(z) = \frac{E(z)(1 - z^{-1})}{T}$$

The differentiation operation is linear, and can be represented by the transfer function $G_d(z)$. We can write

$$G_d(z) = \frac{\text{transform of output}}{\text{transform of input}}$$

Hence

$$G_d(z) = \frac{D(z)}{E(z)}$$
$$= \frac{(1 - z^{-1})}{T}$$

Multiplying numerator and denominator by z gives

$$G_d(z) = \frac{(z - 1)}{Tz}$$

Discrete integration

A very similar procedure can be followed for a discrete approximation to integration, as illustrated in Fig. 9.11. The continuous integral $y(t)$ with respect to time of an error signal $e(t)$

$$y(t) = \int_0^t e(t)\,dt$$

corresponds to the area under the curve of part (a) of the figure. One way of approximating this area, given samples e_n of $e(t)$, is to sum the areas of the rectangular strips shown in Fig. 9.11(b). The approximation y_n to the integral at time nT is then given by the expression

$$y_n = \sum_{i=0}^{n} e_i T$$

Clearly, the smaller the sampling interval T, the closer this approximation is to the value of the integral.

A difference equation representing this form of numerical approximation may be obtained by noting that

$$y_n = y_{n-1} + e_n T$$

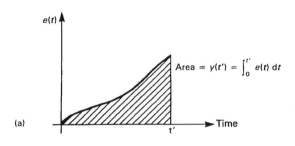

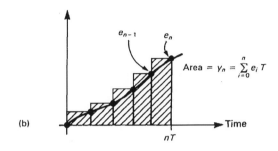

(a)

(b)

Fig. 9.11 Discrete integration.

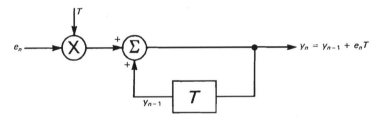

Fig. 9.12 A functional block diagram of discrete integration.

That is, at each sampling instant the new value of the integral is calculated by adding the area of the new strip $e_n T$ to the previous value of the integral y_{n-1}. The corresponding block diagram for the discrete integrator is shown in Fig. 9.12.

Derive a transfer function model of the discrete integrator. **Worked Example 9.7**

Solution Converting the difference equation to sequences we have

$$y[n] = y[n - 1] + e[n]T$$

Transforming gives

$$Y(z) = z^{-1}Y(z) + TE(z)$$

The transfer function of the integrator $G_i(z) = Y(z)/E(z)$. Hence

Note that these particular discrete approximations to integration and differentiation are inverse operations – that is, $G_i(z) = 1/G_d(z)$. This is what might have been expected, given the inverse relationship of ideal differentiation and integation. The discrete approximations described here are not the only ones used, however, as will be seen in Chapter 11.

$$G_i(z) = \frac{T}{(1 - z^{-1})}$$

or, multiplying numerator and denominator by z,

$$G_i(z) = \frac{Tz}{(z - 1)}$$

Worked Example 9.8 A processing operation often used to 'smooth', or low-pass filter, noisy signals is the averaging operation. A two-term moving averager can be defined by the difference equation

$$y_n = \frac{(x_n + x_{n-1})}{2}$$

Draw a block diagram of this averager and derive its transfer function.

Solution The block diagram is shown in Fig. 9.13. To derive the transfer function, convert the difference equation into sequences

$$y[n] = \frac{(x[n] + x[n - 1])}{2}$$

and transform

$$Y(x) = X(z)\frac{(1 + z^{-1})}{2}$$

Hence the transfer function is given by the expression

$$\frac{Y(z)}{X(z)} = \frac{(1 + z^{-1})}{2} = \frac{(z + 1)}{2z}$$

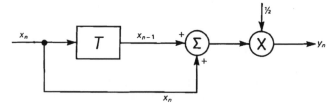

Fig. 9.13 A functional block diagram of a two-term averager.

The z-plane

In the analysis of continuous linear systems, one of the major advantages of the Laplace transform was that it led to s-plane pole–zero models which were easy to interpret in terms of frequency response, transient response and system stability. The z-transform leads to very similar system models. For example, the transfer functions derived in the previous section for the differentiator, integrator and averager can be characterized by pole and zero positions, just like transfer functions in s.

Worked Example 9.9

What are the pole and zero locations of the differentiator, integrator and two-term averager of the previous section?

Solution Recall that zeros are located at values of z which make the transfer function numerator equal to zero, and poles at values which make the denominator zero. Hence

the differentiator, $G_d(z) = \dfrac{(z-1)}{Tz}$, has a pole at $z = 0$ and a zero at $z = 1$

the integator, $G_i(z) = \dfrac{Tz}{(z-1)}$, has a pole at $z = 1$ and a zero at $z = 0$

the averager, $G_a(z) = \dfrac{(z+1)}{2z}$, has a pole at $z = 0$ and a zero at $z = -1$

The previous example involved transfer functions with real poles and zeros only. In general, however, transfer functions in z (like those in s) are ratios of numerator and denominator polynomials which may have real and/or complex roots. So just like Laplace transfer functions, the poles and zeros of a transfer function in z are plotted on a two-dimensional diagram: the z-plane. Figure 9.14 shows the z-plane pole–zero plots of the transfer functions considered so far.

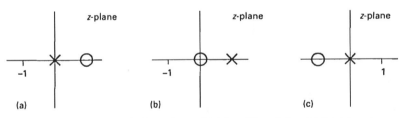

(a) (b) (c)

Note that as with the s-plane, z-plane pole–zero plots give no information about constant multipliers, or 'gain' terms, in the transfer function.

Fig. 9.14 Pole–zero plots in the z-plane of (a) the differentiator; (b) the integrator; and (c) the averager.

Worked Example 9.10

Plot the z-plane pole–zero configuration of the discrete linear system whose input sequence $x[n]$ and output sequence $y[n]$ are related by the difference equation

$$y[n] = x[n] + y[n-1] - 0.5y[n-2]$$

Solution Transforming, we have

$$Y(z) = X(z) + z^{-1}Y(z) - 0.5z^{-2}Y(z)$$

or

$$(1 - z^{-1} + 0.5z^{-2})Y(z) = X(z)$$

189

Hence the transfer function can be written

$$\frac{Y(z)}{X(z)} = \frac{1}{(1 - z^{-1} + 0.5z^{-2})}$$

To obtain a transfer function in powers of z, rather than z^{-1}, multiply numerato and denominator by z^2. Hence we obtain the transfer function

$$\frac{Y(z)}{X(z)} = \frac{z^2}{(z^2 - z + 0.5)}$$

The system therefore has a double zero at $z = 0$ and poles where

$$z = 0.5 \pm j0.5$$

This pole–zero configuration is plotted in Fig. 9.15.

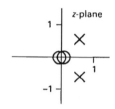

Fig. 9.15   The z-plane pole–zero plot of a second-order discrete system.

Transient response and the z-plane

Look back, if necessary, at the discussion on pages 92–3.

Another analogy between the s-plane and the z-plane is that in both cases th position of the poles can be interpreted directly in terms of system transien response. The argument for transfer functions in z exactly parallels that given i Chapter 5 for those in s. Whatever the nature of the system input, the output ca be manipulated into the sum of a steady-state response and a transient response The latter, whose general form is determined by the system poles, dies away wit time for a stable system. For a discrete system, both steady-state and transien components are, of course, *sequences* rather than continuous waveforms.

Z-plane pole positions must be interpreted differently from s-plane positions however. In order to explain the general implications of z-plane pole positions, technique will first be presented which enables *any* z-transform, expressed as th ratio of two polynomials, to be converted into a corresponding sample sequence

First-order transient response

Figure 9.16 shows the block diagram of a simple system which is used as building block in many digital processors. Its difference equation can be writte immediately as

$$y[n] = x[n] + ay[n - 1]$$

190

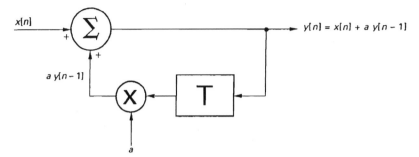

Fig. 9.16 A simple, first-order recursive system.

This equation, involving only sequences and the same sequences shifted by *one* sampling interval, is termed a *first-order* difference equation. Its corresponding (first-order) transfer function is thus

$$G(z) = \frac{Y(z)}{X(z)} = \frac{z}{(z - a)}$$

with a zero at $z = 0$ and a pole at $z = a$.

Now, it was noted earlier that the transform of the unit sample response sequence of a system is identical to the transfer function. So the sequence corresponding to the z-transform

$$G(z) = \frac{z}{(z - a)}$$

is the unit sample response sequence of the system in Fig. 9.16. This sequence can be calculated easily by setting up a table to keep track of what happens after a single unit sample is applied to the system. This is shown in Table 9.2.

The first column, reading downwards, represents the input unit sample sequence

$$x[n] = 1, 0, 0, 0, \ldots$$

This, like the discrete integrator of Fig. 9.12, is a simple example of a *recursive* system, in which the use of feedback means that the output depends on *previous* output values. The discrete differentiator and averager were examples of *non*-recursive systems.

This important technique will recur many times in what follows.

Table 9.2

Input $x[n]$	Delayed output $y[n - 1]$	Output $y[n] = x[n] + ay[n - 1]$
1	0	1
0	1	a
0	a	a^2
0	a^2	a^3
0	a^3	a^4
.	.	.
.	.	.
.	.	.

Assuming, as always, the quiescent condition before the application of the input – that is, there is no 'memory' of any earlier behaviour stored in the delay element – the first row can be filled in immediately. There is 0 in the $y[n-1]$ column and, from the difference equation (or inspection of the block diagram), 1 in the $y[n]$ column.

There are other ways of obtaining this result. The transform $1/(1 - az^{-1})$ could have been expanded as a geometric series (reversing the process for *summing* such a series used earlier in this chapter) or by using the binomial theorem. The alternative, iterative technique presented here, however, is much more widely applicable.

Moving to the second row, the $y[n-1]$ entry can be carried diagonally across and down from row one – it represents, after all, $y[n]$ delayed by one sampling interval. From the difference equation (or directly from the block diagram), the new entry in the output column is a. Proceeding in the same way, the output sequence can be continued indefinitely, obtaining each new term by multiplying the previous term by a. In this way we can identify the z-transform pair

$$1, a, a^2, a^3, a^4, \ldots \quad \leftrightarrow \quad \frac{z}{z-a}$$

Worked Example 9.11 Write down the first few terms of the sequences corresponding to the following transforms:

(i) $z/(z - 1.5)$ (iv) $z/(z + 0.5)$
(ii) $z/(z - 1)$ (v) $z/(z + 1)$
(iii) $z/(z - 0.9)$ (vi) $z/(z + 2)$

Solution The sequences follow immediately from the previous discussion:

(i) $1, 1.5, 2.25, 3.375, \ldots$ (iv) $1, -0.5, 0.25, -0.125, \ldots$
(ii) $1, 1, 1, 1, 1, \ldots$ (v) $1, -1, 1, -1, 1, -1, \ldots$
(iii) $1, 0.9, 0.81, 0.729, \ldots$ (vi) $1, -2, 4, -8, 16, -32, \ldots$

In fact, we can interpret the unit sample response sequence

$$1, a, a^2, a^3, a^4, \ldots$$

as the general form of the *transient response sequence* associated with a single, real pole at $z = a$. Characteristic features of such transient response sequences for real poles are illustrated in Fig. 9.17. They can be summarized as follows:

(1) Providing that $-1 < a < +1$ – that is, the real pole is confined to the region $-1 < z < +1$ in the z-plane – the transient response sequence will die away with time, and the system is stable. Conversely, for real poles located where $|z| > 1$, the system is unstable. Discrete systems with poles at $z = +1$ or -1 are thus on the stability borderline and are associated with a sequence which neither increases nor decreases with time.

Note that the discrete integrator, like its continuous counterpart, has a pole on the stability borderline.

(2) For a real pole lying in the region $0 < z < 1$ the transient response sequence can be thought of as samples of a decaying exponential. The transient response of such discrete systems is non-oscillatory.

(3) For a real pole in the region $-1 < z < 0$ the transient response sequence is oscillatory, with alternate positive and negative values: it can be thought of as samples of a decaying sinusoid at half the sampling frequency.

In addition to the real pole, the system of Fig. 9.16 has a zero at the origin. Poles and zeros at the origin of the z-plane are immaterial, however, for the general

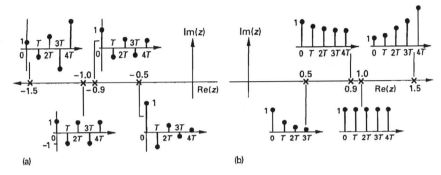

Fig. 9.17   Transient response sequences associated with (a) negative and (b) positive, real, z-plane poles.

form of the transient response sequence. Multiplying a transform by z^{-1} (adding a *pole* at the origin) corresponds to a *delay* of one sampling interval; multiplying by z (adding a *zero* at the origin) corresponds to a time *advance* of one interval. The general form of the transient response term stays the same, but is shifted in time.

Worked Example 9.12

Show, using Table 9.1, that the unit sample response sequence of the processor of Fig. 9.16, with $a = 0.5$, can be thought of as samples of a decaying exponential with a time constant of 1.4 s sampled at 1 Hz.

Solution   The transforms $z/(z - 0.5)$ of the unit sample response sequence, and $z/(z - e^{-aT})$ of the sampled exponential are equivalent providing $e^{-aT} = 0.5$. Hence $-aT = \ln 0.5$ and, since $T = 1$, $a = -\ln 0.5 = 0.69$. The time constant is the reciprocal of this, or approximately 1.4 s.

Second-order transient response

Figure 9.18 shows a simple, *second-order* recursive system. Like the first-order system of Fig. 9.16, systems of this type are used as basic building blocks for a

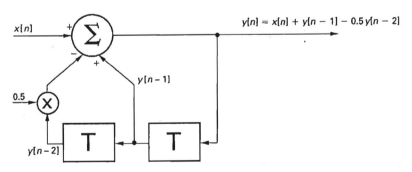

Fig. 9.18   A simple, second-order recursive system.

193

For an introduction to the topic of digital filters see Meade and Dillon (1991).

variety of discrete processors used in digital controllers and filters.

From the figure the difference equation is

$$y[n] = x[n] + y[n - 1] - 0.5y[n - 2]$$

and from Worked Example 9.10 the corresponding transfer function is

$$G(z) = \frac{z^2}{(z^2 - z + 0.5)}$$

with a complex conjugate pair of poles at $z = 0.5 \pm j0.5$.

We can apply exactly the same procedure as in the first-order case and evaluat the unit sample response sequence of the system to obtain the general form of th transient response for these complex poles. The result is shown in Table 9.3: thi time an extra column is needed for $y[n - 2]$ – the output sequence delayed by tw sampling intervals.

Table 9.3

$x[n]$	$y[n - 2]$	$y[n - 1]$	$y[n] = x[n] + y[n - 1] - 0.5y[n - 2$
1	0	0	1
0	0	1	1
0	1	1	0.5
0	1	0.5	0
0	0.5	0	−0.25
0	0	−0.25	−0.25
0	−0.25	−0.25	−0.125
0	−0.25	−0.125	0
.	.	.	.
.	.	.	.
.	.	.	.

The transient response sequence corresponding to a complex conjugate pair o poles at $z = 0.5 \pm j0.5$ is plotted as Fig. 9.19. It can be viewed as samples of decaying sinusoid with a frequency one-eighth of the sampling frequency – that is a decaying sinusoid with eight samples per cycle.

By repeating this process with other second-order discrete processors, witl different multiplier coefficients, it is possible to associate transient respons

Fig. 9.19 Unit sample response sequence of the system of Fig. 9.18.

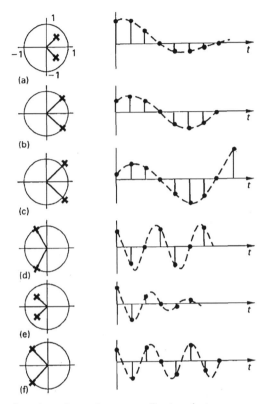

Fig. 9.20 Complex poles and corresponding transient response sequences.

sequences with complex poles in various regions of the z-plane. Typical sequences are shown in Fig. 9.20. (The significance of the circle in the z-plane will become clear in a moment.)

The iterative procedure just introduced is particularly amenable to computer implementation and gives a good insight into the behaviour of such discrete processors. To make a *general* link between complex z-plane poles and transient response, though, we need to proceed analytically.

The use of a spreadsheet for this purpose will be described in Chapter 12.

The positions of a conjugate pair of poles can be conveniently denoted in the polar form

$$z = ce^{\pm j\theta}$$

From Table 9.1, it can be shown (after some manipulation) that the z-transform of the sampled, exponentially-decaying sinusoid $e^{-\sigma t}\sin \omega t$ can be written

$$\frac{Az}{(z - ce^{j\theta})(z + ce^{-j\theta})}$$

where $c = e^{-\sigma T}$, $\theta = \omega T$ and A is a constant determined by σ, T and ω. Now, in one sampling interval, an exponential envelope with a decay factor σ changes by a factor $e^{-\sigma T}$. In the same interval the phase of the sinusoid changes by $\omega T = \theta$ radians. Two important conclusions follow from this.

(1) The distance of the complex z-plane poles from the origin, $c = e^{-\sigma T}$, represents the factor by which the 'envelope' of the response sequence changes in one sampling interval.

(2) If there is a phase change of θ during one sampling interval, then since there are 2π radians in a cycle, there are $2\pi/\theta$ samples per complete period of the oscillatory transient response sequence. In other words, the oscillation has a frequency equal to $\theta/2\pi$ times the sampling frequency.

Check that you understand why the sequences associated with the poles of Fig. 9.20 have the form they do, given their z-plane locations.

So in the second-order example of Table 9.3 and Fig. 9.19, with poles at $z = 0.5 \pm j0.5$, the value of $c = 0.71$ and the decaying 'envelope' changes by this factor each sampling interval. θ in this case is 45°: this is easiest to interpret by noting that 45° is one-*eighth* of a revolution. Hence there are *eight* samples per period of the oscillatory transient response sequence, and the frequency associated with the sequence is one-*eighth* of the sampling frequency.

Note that the z-plane has to be interpreted in the light of a given sampling frequency. Without knowing the sampling frequency we cannot attach any real-time significance to the sequences associated with given pole positions. The z-plane is, if you like, normalized with respect to the sampling frequency. Z-plane pole positions give the relationship between samples in a sequence, but we cannot attach a time-scale to the sequence without knowing f_s, or T.

Worked Example 9.13   The poles in Fig. 9.21 are associated with the z-transform of the sequence obtained by sampling a decaying sinewave of the form $e^{-\sigma t} \sin \omega t$. Estimate the frequency in Hz of the sinusoidal part of the oscillation if the sampling frequency is 12 Hz. By what factor will the envelope decay in 0.5 s?

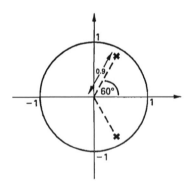

Fig. 9.21   A pair of complex poles.

Solution   The poles lie at an angle of 60°, or $\pi/3$ rad. This corresponds to $\frac{1}{6}$ of a revolution and the associated frequency is therefore $12/6 = 2$ Hz.

The poles are 0.9 units from the origin of the z-plane. A time interval of 0.5 s corresponds to 6 sampling intervals, so the envelope of the exponentially decaying sinewave will decay by a factor of $0.9^6 = 0.53$, that is to about half of its initial value.

Stability of discrete linear systems

Understanding the relationship between the values of c and θ, for poles at $z = ce^{\pm j\theta}$, and the corresponding sequences is the key to interpreting the z-plane. The sequences shown in Fig. 9.20 illustrate the relationship between pole angle and the number of samples per cycle of oscillation. As the pole angle gets bigger and the pair of poles moves from the right-hand to the left-hand side of the imaginary axis in the z-plane there are fewer samples per cycle.

The value of c determines whether the sequence grows or decays with time. If c is less than 1 then the poles will lie inside a circle of radius 1 centred on the origin of the z-plane. This is called the *unit circle* and it is usual to include it on z-plane pole–zero diagrams. Any pole located within the unit circle is associated with a sequence that dies away with time. The smaller the value of c the closer the poles will lie to the origin of the z-plane and the faster the sequence will decay, relative to the sampling period.

Poles that lie outside the unit circle, where c is greater than 1, correspond to sequences whose 'envelopes' grow indefinitely with time. Transfer function poles in this region indicate that the corresponding sampled-data system will be unstable. For stability, therefore, all poles associated with the system model must lie within the unit circle.

Poles that lie on the circumference of the unit circle, as in Fig. 9.20(b) and (f), correspond to sequences that oscillate steadily with time. In this case $c = 1$ and the poles lie at $z = e^{\pm j\theta}$. The ratio $\theta/2\pi$ gives the relationship between the oscillation frequency and the sampling frequency.

A note of caution

It is convenient to interpret z-plane pole positions as corresponding to samples of the exponentially decaying sinusoids which are so useful in our *continuous* models of signals and systems. These interpretations form an important link between discrete and continuous models, and will be developed further in the next chapter. It is equally important, however, always to bear in mind that the z-transform is not a unique representation of anything other than a particular sequence of numbers. A z-plane model of a sample sequence gives no information whatsoever about the behaviour between sampling instants of the continuous signal from which it was derived. Different continuous signals which produce the same set of sample values will give rise to identical z-transforms. This means that given only the z-transform of a sequence of samples we cannot (without additional information) reconstruct the waveform of the underlying continuous signal. The dotted 'envelopes' of Fig. 9.20 should therefore be viewed with caution.

In the light of this caveat, we can interpret the phenomenon of aliasing on the z-plane. Recall from the beginning of the chapter that half the sampling frequency $f_s/2$ is the maximum frequency that can be represented unambiguously in sampled form. This fact is reflected in our interpretation of the z-plane. Suppose that conjugate poles exist on the unit circle at $z = e^{\pm j\pi/4}$. If the sampling rule has been followed, these can represent a sinewave of frequency f sampled at $f_s = 8f$, since we have the relationship $f/f_s = \theta/2\pi$.

Suppose now, however, that the sampling rule has not been followed and the sinewave was sampled less than once per cycle, at $f_s = 8f/9$. Then $2\pi f/f_s = \theta = 18\pi/8$, or $2\pi + \pi/4$. Since we cannot distinguish between an angle of $\pi/4$ and an angle of $2\pi + \pi/4$ on the z-plane, the poles corresponding to the undersampled sinewave will lie at exactly the same position as those of the correctly sampled signal.

This is simply another manifestation of the fact that the z-plane deals only with sequences and contains no information about what happens between sampling instants. A given set of samples may equally well come from a higher frequency sinewave sampled too slowly, or indeed an arbitrary signal with higher frequency components which just happens to produce the same set of sample values.

Undersampling results in aliasing and in the z-plane the poles of an aliased signal cannot be distinguished from those of a correctly sampled signal. In digital control systems design it is very much up to the engineer to ensure that sampling rates are high enough for the sampled output of a controlled continuous system to be an accurate reflection of the behaviour of the system. If aliasing occurs then z plane models will not provide information about what is actually happening to the system.

Summary

Digital control systems involve the sampling of continuous signals. In order to avoid aliasing, a continuous signal with components from 0 to B Hz must be sampled at a rate greater than $2B$ samples per second.

Samples are often processed by computer algorithms which can be modelled as discrete linear systems – that is, the algorithms operate on an input sample sequence to produce an output sequence, obeying the principle of superposition. Such linear algorithms include discrete approximations to differentiation and integration, and moving averagers used to 'smooth' noisy sequences.

The z-transform converts a sequence of samples into a polynomial in powers of z^{-1}, where z^{-1} may be thought of as a delay operator. The transfer function of a discrete linear system can then be defined as

$$\frac{z\text{-transform of output sequence}}{z\text{-transform of input sequence}}$$

assuming quiescence before applying the input. Such transfer functions in z may be derived directly from the system difference equation.

Stable discrete linear systems have poles located within the unit circle in the z plane. Transient response sequences can be associated with system pole positions. A complex conjugate pair of poles within the unit circle is associated with an oscillatory, decaying sequence. The rate of decay is determined by the distance of the poles from the origin (modulus), while the associated frequency of oscillation is determined by their angle (argument). Poles on the real axis may be thought of as limiting cases: the region between $z = 0$ and $z = 1$ (angle = zero) is associated with a non-oscillatory, decaying sequence; the region between 0 and −1 (angle = π radians) is associated with a decaying oscillatory sequence at half the sampling frequency.

Problems

9.1 A signal can be modelled by the expression $y(t) = 4t + 2$. It is sampled twice per second, starting at time $t = 0$. What is the z-transform of the first four sample values?

9.2 What is the z-transform $F(z)$ of the signal $f(t) = 1 - \exp(-2t)$, sampled at a frequency of $10\,\text{Hz}$?

9.3 An exponentially increasing signal has the Laplace transform $F(s) = 2/(s - 3)$. What is its z-transform when sampled once per second?

9.4 The unit sample response sequence of a processor is

$$0, 2, 2.5, 0.5, 0, 0, 0, \ldots$$

Sketch the pole–zero diagram of the processor.

9.5 One version of smoothed digital derivative action is defined by the difference equation

$$d_n = \frac{1}{6T}[e_n + 3e_{n-1} - 3e_{n-2} - e_{n-3}]$$

By arranging three delay blocks as a 'tapped delay line', so that the signals *between* the delay blocks are also available for processing, draw a block diagram representing this algorithm. What is the transfer function of the differentiator?

9.6 The form of digital integration described in this chapter is often known as *backward integration*. An alternative, known as *forward integration*, is illustrated in Fig. 9.22.
(a) Express forward integration as a difference equation.
(b) Draw the block diagram corresponding to forward integration.
(c) What is the transfer function of this discrete integrator?

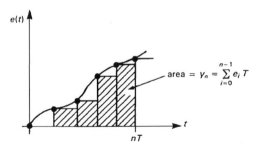

Fig. 9.22 Forward integration.

9.7 A system is modelled by a transfer function with poles at $z = 0.3 \pm j0.4$ when the sampling frequency is 5 Hz.
(a) What is the frequency of the transient oscillations of the system?
(b) How long does it take for the envelope of the transient oscillations to decay to one-eighth of its initial value?

9.8 A digital processor used as a compensator has the transfer function

$$\frac{z - 0.4}{z - 0.6}$$

What is the difference equation relating the input sequence $x[n]$ to the output sequence $y[n]$?

Digital control II. Sampled-data systems   10

□ To show how the discrete models of Chapter 9 can be applied to sampled-data systems.   **Objectives**
□ To explain how open- and closed-loop transfer functions in z can be derived for digital control systems.
□ To model the effect of changing gain and sampling rate on the closed-loop behaviour of simple digital control systems.
□ To show how the root-locus technique can be used in the z-plane.
□ To explain the link between the s- and z-planes.

Figure 10.1(a) shows the basic block diagram of a digital closed-loop control system in unity feedback form. Systems like this, which involve the discrete processing of samples of continuous signals are often known as *sampled-data systems*. Figure 10.1(b) is a conceptual version of the system, showing the functional characteristics of each element in the system and the nature of the signals between the various blocks. The ADC is represented by a sampler, and another sampler is shown at the system input as a reminder that the error calculation is performed on discrete values. The DAC is modelled as an element which responds to each input number by producing a steady output held at a constant value for one sampling period. This model is known as a zero-order hold, because the DAC is said to interpolate a zero-order polynomial – that is, a constant value – between samples. Higher-order data interpolation is rare in control engineering, so in this text the simple term 'hold' will be used to refer to this zero-order model.

The digital controller in Fig. 10.1(b) processes the input error sequence to produce an output sequence. As described in the previous chapter, providing the operations carried out by the controller are linear, such a controller can be modelled by a linear difference equation and, equivalently, a transfer function in z. Specific examples will be discussed in the next chapter. Here we turn first to the derivation of a transfer function in z of the remainder of the system, and then to discrete models of complete closed-loop behaviour.

Data conversion normally introduces a scaling factor or constant gain: for example, the binary word 0000 0001 applied to an 8-bit DAC is unlikely to give an output of 1 volt! For modelling purposes we shall assume that gains associated with data conversion are included in the overall loop gain of the system model.

The pulse transfer function

The fundamental step in the analysis is to note that we can consider the combination of hold, plant and output sampler as a discrete system in its own right. Input samples (from the controller algorithm) are processed to give output samples (from the ADC). So once again, providing the plant can be modelled as a linear system, the hold, plant and output sampler can be characterized by a linear difference equation or a transfer function in z. Remember, though, that such a

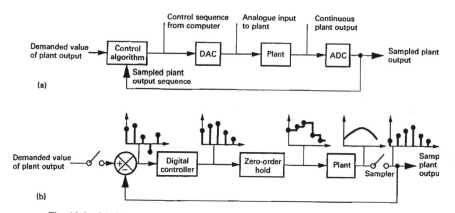

Control sequence from computer

Analogue input to plant

Continuous plant output

Demanded value of plant output

Control algorithm

DAC

Plant

ADC

Sampled plant output

Sampled plant output sequence

(a)

Demanded value of plant output

Digital controller

Zero-order hold

Plant

Sampler

Samp plant outpu

(b)

Fig. 10.1   (a) Components of a digital control system. (b) The signals in the system.

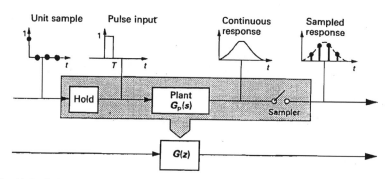

Unit sample

Pulse input

Continuous response

Sampled response

Hold

Plant $G_p(s)$

Sampler

$G(z)$

Fig. 10.2   Definition of the pulse transfer function of the plant combined with a hold and sampler.

This is an extremely important point to bear in mind whenever such techniques are used in control system analysis and design. As part of the design process it is good practice to simulate the sampled-data system to ensure that behaviour between samples is acceptable.

discrete model will give no information about the plant output between samplin instants.

We can proceed directly from the plant transfer function $G_p(s)$ to the *puls transfer function* $G(z)$ of the hold, plant and sampler combined. Figure 10.2 show how this is done. Recall that the transfer function of a discrete system is equal t the transform of the unit sample response. A unit sample input to the hold gives unit output held steady for one sampling interval – that is, a rectangular pulse c unit height and width T as shown in Fig. 10.2. The continuous response of th plant to this pulse is sampled to give the output sequence. And the z-transform c this sequence is, by definition, the pulse transfer function $G(z)$.

It is not necessary to perform such a unit sample response test: the technique i simply a way of manipulating our models into the required form. The Laplac transform is first used to work out the (continuous) response of the plant to th rectangular pulse. Standard tables like Table 9.1 are then used to derived the z transform of the corresponding sampled response.

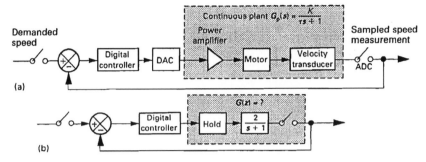

Fig. 10.3 Motor speed control loop.

Consider, for example, the simple motor speed control loop of Fig. 10.3(a). Modelling the amplifier and velocity transducer as pure gains, and the motor as a first-order lag, we can write

$$G_p(s) = \frac{K}{(1 + s\tau)}$$

where K is the overall low-frequency gain and τ is the motor time constant. Assuming values of $K = 2$ and $\tau = 1$, the conceptual block diagram of the closed-loop system can be drawn as Fig. 10.3(b).

The hold produces a rectangular pulse in response to a unit sample input. Figure 10.4 shows how this pulse can be represented as a step up at time $t = 0$ followed by a step down at $t = T$. The response of the plant can therefore be represented as the combination of a positive unit step response starting at time $T = 0$ and a negative unit step response at time $t = T$. $G(z)$, the pulse transfer

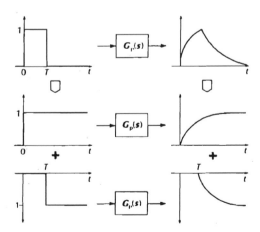

Fig. 10.4 Pulse response as the sum of positive-going and negative-going step responses.

function of the hold, plant and sampler combined, is simply the z-transform of the sampled version of this combined response.

The Laplace transform of a positive unit step is $1/s$, while that of the delayed negative step is $-e^{-sT}/s$. The transform of the positive step response is therefore $G_p(s)/s$, and that of the delayed negative step response is $-e^{-sT}G_p(s)/s$. The transform of the total rectangular pulse response is hence the sum

$$(1 - e^{-sT})\, \frac{G_p(s)}{s}$$

Similarly the *sampled* rectangular pulse response can be thought of as the sum of the *sampled* 'step up' and 'step down' parts, as shown in Fig. 10.5. Let us denote the z-transform of the response to the 'step up' by

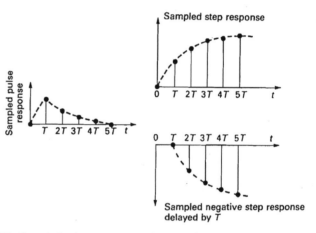

Fig. 10.5 Sampled pulse response as the sum of two sampled step responses.

This can be interpreted as: 'look up the z-transform of the sampled signal whose continuous form has the Laplace transform $G_p(s)/s$'.

$$Z\left[\frac{G_p(s)}{s}\right]$$

Since a delay of one sampling interval, T seconds, corresponds to multiplication by the factor z^{-1}, the delayed 'step-down' sampled response has the z-transform

$$-z^{-1}\, Z\left[\frac{G_p(s)}{s}\right]$$

So the complete, sampled response, equivalent to the pulse transfer function of the hold–plant–sampler combination, can be written as the sum

$$G(z) = Z\left[\frac{G_p(s)}{s}\right] - z^{-1}\, Z\left[\frac{G_p(s)}{s}\right] = (1 - z^{-1})\, Z\left[\frac{G_p(s)}{s}\right]$$

Multiplying numerator and denominator by z leads to the conventional form

$$G(z) = \frac{(z-1)}{z} \, Z\left[\frac{G_p(s)}{s}\right]$$

What is the pulse transfer function $G(z)$ for the motor system of Fig. 10.3(b):

(a) for a general sampling interval T?
(b) for a sampling rate of 20 Hz?

Solution
(a) From Fig. 10.3(b)

$$G_p(s) = \frac{2}{(s+1)}$$

so we need first to find the z-transform corresponding to

$$\frac{G_p(s)}{s} = \frac{2}{s(s+1)}$$

Table 9.1 gives the correspondence

$$Z\left[\frac{1}{s(s+a)}\right] \quad \text{---sample}\longrightarrow \quad \frac{z(1-e^{-aT})}{(z-1)(z-e^{-aT})}$$

If the Laplace transform is scaled by a constant K, then so is the z-transform of the corresponding sampled signal. In this case $a = 1$ and, scaling by a factor 2, we have

$$Z\left[\frac{2}{s(s+1)}\right] \quad \text{---sample}\longrightarrow \quad \frac{2z(1-e^{-T})}{(z-1)(z-e^{-T})}$$

Substituting in the expression of the pulse transfer function

$$G(z) = \frac{(z-1)}{z} \, Z\left[\frac{G_p(s)}{s}\right]$$

gives

$$G(z) = \frac{(z-1)}{z} \, \frac{2z(1-e^{-T})}{(z-1)(z-e^{-T})}$$

or

$$G(z) = \frac{2(1-e^{-T})}{(z-e^{-T})}$$

(b) For a sampling rate of 20 Hz, $T = 1/20 = 0.05$. Substituting this value leads to

$$G(z) = \frac{0.1}{(z-0.95)}$$

To save tedious algebraic manipulation, reference books often tabulate pulse transfer functions for common plant models combined with hold and sampler. Computer software, too, is widely used for such purposes. Table 10.1 lists some examples for information.

Although control system design software has eliminated much of the need for engineers to carry out the algebraic manipulations by hand, it is still important to understand the ideas behind the models.

Table 10.1

Plant transfer function $G_p(s)$	Pulse transfer function $G(z)$ when combined with hold and sampler
$\dfrac{1}{s}$	$\dfrac{T}{z-1}$
$\dfrac{1}{s^2}$	$\dfrac{T^2(z+1)}{2(z-1)^2}$
$\dfrac{1}{s+a}$	$\dfrac{1-e^{-aT}}{a(z-e^{-aT})}$
$\dfrac{1}{s(s+a)}$	$\left[\dfrac{aT-1+e^{-aT}}{a^2}\right]\dfrac{(z+b)}{(z-1)(z-e^{-aT})}$ $\left(\text{where } b = \dfrac{1-e^{-aT}-aTe^{-aT}}{aT-1+e^{-aT}}\right)$

The closed-loop transfer function

Figure 10.6(a) shows an example of a temperature control loop using a digital controller. The controlled temperature is a continuous variable, and a sampled version of this is used as an input to the control algorithm. The system is redrawn in part (b) of the figure: $G_p(s)$ represents the combined transfer function of the actuator, valve and reaction vessel, and the ADC is modelled, as usual, as a sampler. Note, however, that *fictitious samplers* have been added at the input and

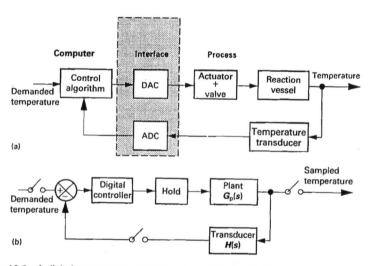

Fig. 10.6 A digital temperature-control system. Note the 'fictitious samplers' in part (b).

output of the loop. These do not correspond to any physical device (and do not affect system behaviour) but have two important functions. First, as noted in the introduction to this chapter, they serve as a reminder that discrete models in z deal with input and output *samples* only; furthermore, they enable a closed-loop transfer function to be derived in a way very similar to the closed-loop transfer functions in s or $j\omega$ used earlier in this book. Recall that in each case, the closed-loop transfer function took the form

$$\frac{\text{forward path transfer function}}{1 + \text{open-loop transfer function}}$$

The same approach can be applied, with a little care, to a sampled-data system, as shown in Fig. 10.7. The previous section showed how to derive the transfer function $G(z)$ illustrated in part (a): the forward path transfer function is then simply $KG(z)$. The open-loop transfer function can be obtained in a similar manner, but it is now necessary to consider the combination of hold, $G_p(s)$, $H(s)$ and sampler, as illustrated in Fig. 10.7(b). This combination has the pulse transfer function

$$\frac{(z-1)}{z} Z\left[\frac{G_p(s)H(s)}{s}\right]$$

usually written as $GH(z)$. The open-loop pulse transfer function is therefore $KGH(z)$ and the closed-loop pulse transfer function is hence

$$\frac{KG(z)}{1 + KGH(z)}$$

In the case of complex sampled-data systems – for example, systems with multiple loops, and data converters in various positions – great care is required in

The notation $GH(z)$ is a reminder that the continuous transfer functions $G_p(s)$ and $H(s)$ must be combined *before* converting to a z-transform. Only if the process and transducer were separated by a sampler and hold would the pulse transfer function of their combination be $G(z)H(z)$. In general, $G(z)H(z) \neq GH(z)$.

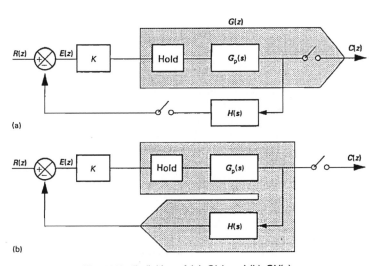

(a)

(b)

Fig. 10.7 Definition of (a) $G(z)$ and (b) $GH(z)$.

carrying out the manipulations necessary to derive a pulse transfer function
Indeed, certain sampled-data topologies (beyond the scope of this book) do no
even possess a pulse transfer function!

Worked Example 10.2 Figure 10.8 shows the motor speed control example introduced earlier in unit
feedback form with fictitious samplers. Derive the closed-loop pulse transfe
function $C(z)/R(z)$ for a general sampling interval T and controller gain K.

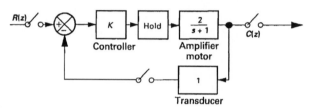

Fig. 10.8 Digital speed control loop.

Solution From Worked Example 10.1 the forward path transfer function is

$$KG(z) = \frac{2K(1 - e^{-T})}{(z - e^{-T})}$$

Since the loop has unity feedback, the closed-loop pulse transfer function can b
written

$$\frac{C(z)}{R(z)} = \frac{KG(z)}{1 + KG(z)} = \frac{2K(1 - e^{-T})}{(z - e^{-T}) + 2K(1 - e^{-T})}$$

Hence

$$\frac{C(z)}{R(z)} = \frac{2K(1 - e^{-T})}{z + [2K(1 - e^{-T}) - e^{-T}]}$$

Once appropriate values are substituted for the gain K and sampling interval T
this expression will simplify to the form $A/(z + a)$, where A and a are constants
In other words, the closed-loop system has one real z-plane pole, whose positioi
is determined by the gain *and* the sampling rate.

Root locus in the z-plane

Given an open-loop transfer function in z, the root-locus technique of Chapter
may be applied in the z-plane to plot all possible closed-loop pole positions a
controller gain (or some other parameter) is varied.

For the configuration of Fig. 10.7, closed-loop poles lie where

$$1 + KGH(z) = 0$$

This equation is similar in form to that used to derive the rules of root-locus construction in Chapter 6. So if the gain K is varied, exactly the same rules can be used for the z-plane locus as for the s-plane. The transient response sequences associated with particular z-plane pole positions must, however, be interpreted in the light of the discussion of Chapter 9. Consider, for example, the motor speed control loop of Fig. 10.8. With a sampling rate of 20 Hz, the open-loop pole lies at $z = 0.95$, as derived in Worked Example 10.1. The corresponding root locus is therefore as in Fig. 10.9. For low values of gain, the closed-loop pole is located near the open-loop location $z = 0.95$, implying a transient response sequence taking many sampling intervals to die away. As the gain is increased, the closed-loop pole moves towards the left and the response sequence dies away faster and faster. If the gain is increased sufficiently for the pole to be located between $z = 0$ and $z = -1$, the transient response is oscillatory, with a frequency equal to half the sampling frequency, and takes more and more sampling intervals to die away as the pole moves closer to the point $z = -1$. If the gain is increased still further, the closed-loop pole lies outside the unit circle in the z-plane, and the system is unstable.

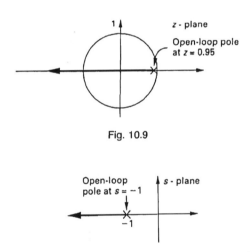

Fig. 10.9

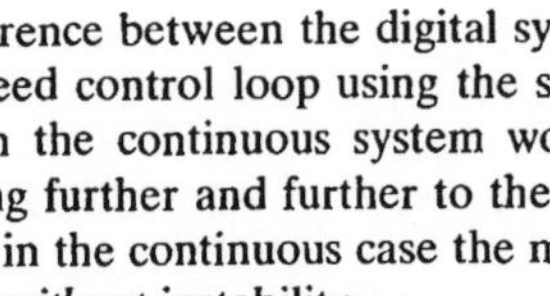

Fig. 10.10 The z-plane root locus of the motor speed control system.

Note a fundamental difference between the digital system and a corresponding continuous proportional speed control loop using the same model of the motor. Increasing the loop gain in the continuous system would result in the s-plane closed-loop pole also moving further and further to the left along the real axis as shown in Figure 10.10. But in the continuous case the model predicts a faster and faster closed-loop response *without* instability.

Note that for one particular value of K, the closed-loop pulse transfer function consists of a single pole at the origin of the z-plane. In other words, the output sample sequence is equivalent to the input sequence delayed by one sampling interval. For an input step, therefore, the output sequence must reach its final

In practice, of course, the modelling assumptions would become invalid if gain were increased too much in the continuous system, so the closed-loop response could not be made faster indefinitely by increasing loop gain. Nevertheless, the distinction between the discrete and continuous models remains an important one.

value in one sampling interval and remain there, as illustrated in Fig. 10.11. This type of response, in which the output sequence settles to its steady-state form after a (small) finite number of sampling intervals, is known as a deadbeat response. Deadbeat response forms the basis of certain design methods for digital control systems – although such methods must be used with a great deal of caution. One problem which can easily arise is that the behaviour of the plant output *between* sample instants can be highly oscillatory and unsatisfactory, even though the output *samples* possess the deadbeat property.

Such intersample behaviour will be considered further in the next chapter.

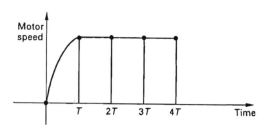

Fig. 10.11   A deadbeat response.

Worked Example 10.3   Derive the closed-loop pulse transfer function for the level control system of Fig. 10.12 with a sampling rate of 1 Hz. Sketch the root locus for varying gain K, and explain how the closed-loop transient response changes as K is steadily increased from zero.

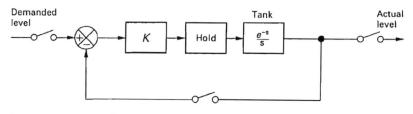

Fig. 10.12   A discrete level control system.

Suppose the gain is increased to the point at which the system is on the verge of instability. What is the frequency of oscillation of the system output?

Note that a delay of any small, whole number of sampling intervals can easily be incorporated into a transfer function in z by adding poles at the origin. This is in stark contrast to delays in continuous systems, which are difficult to deal with in the s-plane.

Solution   The term e^{-s} in the plant transfer function corresponds to a delay of 1 s or one sampling interval. The plant pulse transfer function is therefore equivalent to that of a pure integrator (with hold and output sampler) multiplied by z^{-1} to take into account the delay.

The pulse transfer function of a hold, integrator and sampler is, from Table 10.1, $T/(z - 1)$. For $T = 1$, and incorporating the delay, we have an open-loop pulse transfer function

210

$$KG(z) = \frac{K}{z(z-1)}$$

The root locus is shown in Fig. 10.13. As the gain is increased from zero, the poles at $z = 0$ and $z = 1$ approach each other on the real axis. The transient response sequence consists of two terms, both non-oscillatory, and initially dominated by the term associated with the pole nearer to $z = 1$. As the gain is

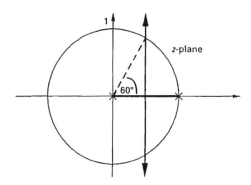

Fig. 10.13   The root-locus plot of the system of Fig. 10.12.

increased, therefore, the transient response at first dies away faster, without oscillation. If K is increased beyond a certain value, however, the closed-loop poles form a complex conjugate pair, corresponding to an oscillatory response. The angle of the poles increases with increasing K, and the frequency of the transient oscillation therefore also increases. The oscillations take more and more sample intervals to die away as the poles move closer to the unit circle. At sufficiently high gain, the closed-loop poles are located outside the unit circle and the system is unstable.

Refer to the transient response sequences described in Chapter 9, particularly Figs 9.17 and 9.20.

On the verge of instability the complex poles have an angle of arccos $0.5 = 60°$. The frequency of oscillation is therefore one-sixth of the sampling frequency – in this case 0.17 Hz.

Varying the sampling rate

In general, the pulse transfer function will depend on T, the sampling interval, as well as on other system parameters. Varying T will therefore affect the closed-loop pole positions, and hence the system response. As an example, we return to the simple motor speed control loop.

Sketch, on numerically labelled axes, the general form of the transient response sequence of the motor speed control loop of Fig. 10.8 if the controller gain is fixed at $K = 9.5$ but the sampling frequency is

Worked Example 10.4

(a) increased to 50 Hz
(b) decreased to 12 Hz.

At what sampling frequency does the system become unstable?

Solution From Worked Example 10.2, the closed-loop pole of the speed contr(loop lies at

$$z = -[2K(1 - e^{-T}) - e^{-T}]$$

(a) For $K = 9.5$ and $T = 1/50 = 0.02$ s, the pole lies at $z = 0.6$. The associate transient response sequence, from Chapter 9, is thus

  1, 0.6, 0.36, 0.22, 0.13, . . .

as plotted in Fig. 10.14(a).

(b) With $K = 9.5$ and $T = 1/12 = 0.083$ s, the closed-loop pole lies at $z = -0.($ and the associated response is

  1, −0.6, +0.36, −0.22, +0.13, . . .

as plotted in Fig. 10.14(b).

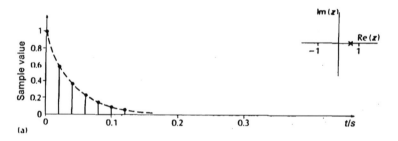

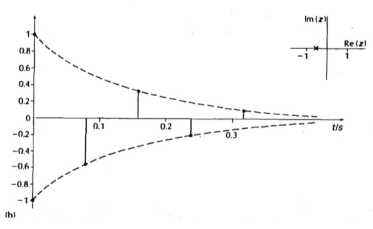

Fig. 10.14

212

Decreasing the sampling frequency, then, moves the pole left along the real axis: onset of instability occurs when $z = -1$. For a gain of 9.5, therefore, the system becomes unstable when

$$-[19(1 - e^{-T}) - e^{-T}] = -1$$

That is, when $e^{-T} = 0.9$ or $T = 0.105\,\text{s}$. This corresponds to a sampling frequency of $1/0.105$ or $9.5\,\text{Hz}$.

Note the difference in time-scale of the two plots of Fig. 10.14, despite the fact that in each case the system pole is the same distance from the origin of the z-plane. As was noted in the previous chapter, the real-time implications of z-plane pole positions can only be properly interpreted in the light of the sampling interval.

Decreasing the sampling rate, then, can result in undesirable transient behaviour or instability. One way of looking at this effect is that too slow a sampling rate is equivalent to introducing a delay into a loop. And as was discussed earlier in this book, excessive time delays can lead to instability because correcting control action cannot be applied sufficiently quickly.

Simulating complete system response

Although a knowledge of the general form of the transient response of a system is a useful aid to design, it is often necessary to compute the complete response of a control system to some test input – a step, for example. The tabular technique introduced in Chapter 9 can be exploited directly in such simulations. For example, consider the motor speed control loop with $K = 4.6$ and $T = 0.05\,\text{s}$ (20 Hz). From Worked Example 10.2, the closed-loop transfer function is

$$\frac{C(z)}{R(z)} = \frac{2K(1 - e^{-T})}{z + [2K(1 - e^{-T}) - e^{-T}]} = \frac{0.45}{z - 0.5}$$

To use the tabular method, the transfer function must be converted to a difference equation involving sequences and delayed sequences. Multiplying numerator and denominator by z^{-1} and cross-multiplying we have

$$(1 - 0.5z^{-1})C(z) = 0.45z^{-1}R(z)$$

Converting to sequences gives

$$c[n] - 0.5c[n - 1] = 0.45r[n - 1]$$

or

$$c[n] = 0.45r[n - 1] + 0.5c[n - 1]$$

To evaluate the step response sample by sample, we set up a table with columns for $c[n]$, $c[n - 1]$, $r[n]$ and $r[n - 1]$ as shown in Table 10.2. The output response $c[n]$ is shown in Fig. 10.15.

The steady-state output value can be derived immediately from the difference equation by noting that in the steady state $c_n = c_{n-1} = c_{ss}$ and $r_n = r_{n-1} = 1$. Hence

213

Table 10.2

$r[n]$	$r[n-1]$	$c[n-1]$	$c[n] = 0.45\, r[n-1] + 0.5c[n-1]$
1	0	0	0
1	1	0	0.45
1	1	0.45	0.675
1	1	0.675	0.7875
1	1	0.7875	0.8437
1	1	0.8437	0.8719
.	.	.	.
.	.	.	.
.	.	.	.

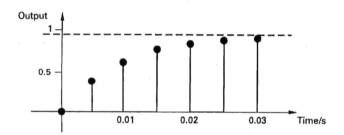

Fig. 10.15   Step response samples of motor speed control system with $K = 4.5$ and a sampling frequency of 20 Hz.

$$c_{ss} = 0.45 + 0.5c_{ss}$$

and

$$c_{ss} = 0.45/0.5 = 0.9$$

which is in accordance with the trend of Fig. 10.15.

The relationship between the s-plane and z-plane

At this point it is as well to formalize the relationship between the s- and z-plane – a relationship which follows, in fact, from the discussion of discrete system transient response in the previous chapter. If we take the Laplace transform of a continuous signal, and the z-transform of the sampled version of the same signal then there is a simple relationship between the poles of the Laplace transform and the poles of the z-transform. If the transform of the continuous signal has a pole at $s = -a$ (where a may be real or complex), the z-transform has a pole at $z = e^{-aT}$. This property also applies to pulse transfer functions, as can be seen by looking back at Table 10.1. If the continuous plant has a pole at $s = -a$, then the pulse transfer function of hold, plant and sampler combined has a pole at $z = e^{-aT}$.

This relationship between s-plane poles and z-plane poles can be written generally as $z = e^{sT}$.

For a particular pole location at $s_0 = \sigma_0 + j\omega_0$ we can write

$$z = e^{(\sigma_0 + j\omega_0)T} = e^{\sigma_0 T} e^{j\omega_0 T}$$

Using the usual polar notation for pole position

$$z = ce^{j\theta}$$

we can identify

$$c = e^{\sigma_0 T} \quad \text{and} \quad \theta = \omega_0 T$$

This is consistent with other aspects of s and z operators. For example, z^{-1} corresponds to a delay of one sampling interval, as does e^{-sT}.

Figure 10.16 illustrates this relationship, and links stability criteria in the s- and z-planes. In the s-plane, the condition for stability is that all poles lie in the left half plane: transient response terms all have negative σ, and die away exponentially with time. In the z-plane, the poles of stable systems must lie within the unit circle: c is less than 1, so the transient envelope gets smaller with each successive sampling interval. Conversely, s-plane poles in the right half plane, or z-plane poles outside the unit circle, indicate instability: both are associated with responses which increase indefinitely with time.

Care is necessary to avoid serious misconceptions when interpreting the relationship between the s-plane and z-plane. The expression $z = e^{sT}$ holds between the s-plane poles of a continuous system model *and the z-plane poles of its pulse transfer function when combined with a hold at the input and a sampler at the output*. It does *not* necessarily hold for the closed-loop s-plane poles of a continuous system and the closed-loop z-plane poles of a discretized version of it.

Second, the relationship holds only for poles, not for zeros. The complete pole–zero picture for a given system can only be found by using the technique explained above for obtaining the complete pulse transfer function. And zeros, of course, can have a profound effect on system behaviour – as was stressed in Chapters 5 and 6.

In spite of these cautionary remarks, the general association $z = e^{sT}$ is useful for design purposes. Indeed, one design method uses the relationship to *define* a digital compensator as the transfer function obtained from an original continuous

The motor speed control system discussed earlier showed this clearly. The model of the continuous, closed-loop system implies stability for all values of gain, with the s-plane pole always in the left half plane. In contrast, the z-plane poles of the digital version can move outside the unit circle if the sampling rate is sufficiently low or the gain sufficiently high – in which case the above mapping of s-plane to z-plane certainly does not hold.

This design technique is known as *pole–zero mapping*.

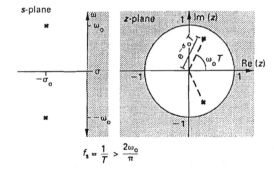

Fig. 10.16 The link between s-plane poles and z-plane poles.

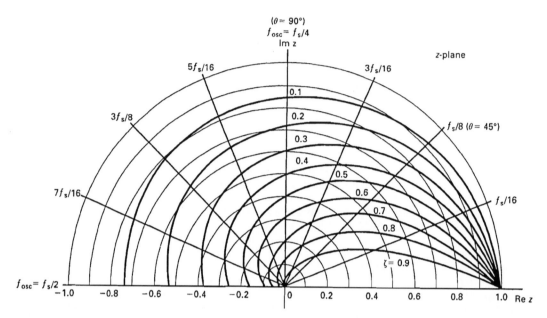

Fig. 10.17   The z-plane, showing the relationship between pole position and damping ratio, decay factor and oscillation frequency.

design in s by applying the relationship $z = e^{sT}$ to s-plane poles *and* zeros. (The dynamic behaviour of such a compensator might differ considerably from its continuous prototype.)

We can also use the expression $z = e^{sT}$ to characterize complex pole positions in the z-plane by means of the concept of damping ratio familiar from our analysis of continuous second-order systems. From Fig. 5.9 it can be seen that continuous systems with a given damping ratio ζ have a constant ratio between the real and imaginary parts of s-plane pole positions. Because of this, the loci of constant damping ratio in the s-plane are the (straight) damping lines of Chapter 5. Similar loci of constant ζ can be constructed in the z-plane but owing to the exponential relationship between s- and z-plane pole positions they are not straight lines. (They can be shown to be logarithmic spirals.) Such lines of constant ζ are illustrated in Fig. 10.17. The figure is also annotated so as to be able to read off the modulus and angle of any given pole, thus enabling the decay factor and oscillation frequency to be estimated easily.

Worked Example 10.5   A second-order digital control system has a zero at the origin of the z-plane and a complex conjugate pair of poles. It uses a sampling rate of 50 samples per second. Step response overshoot is around 10%, and the system settles to 5% of its final value in 0.5 s. Suggest approximate z-plane pole positions based on Fig. 10.17. What would be the frequency of oscillation of the transient response of the system? If possible, compute the system discrete step response for your suggested pole locations.

Solution For a pole where $|z| = c$, the transient response envelope decays by a factor of c in one sampling interval. Here, the envelope must be less than 0.05 of its original value after 25 sampling intervals. Hence c is given by

$$c^{25} = 0.05$$

Taking logarithms gives

$$25 \log c = -1.3 \quad \text{and} \quad c = 0.887$$

Check: $0.887^{25} = 0.499$

Ten percent overshoot corresponds to a damping ratio of 0.6, so from Fig. 10.17, $\theta \approx 10°$. The frequency of the transient oscillation is therefore $50/36 \approx 1.4\,\text{Hz}$. The simulated discrete step response is shown in Fig. 10.18, which is very similar to samples of a continuous second-order step response with damping ratio 0.6.

As always in the z-plane, it must be borne in mind that the models deal with sample sequences only. Unless there is a significant number of samples per cycle of the oscillatory transient (as in the previous worked example), a value of ζ read from Fig. 10.17 will not necessarily lead to accurate predictions of, say, step response peak overshoot. Nevertheless, for design methods using z-plane root locus, and providing that the sampling rate is reasonably high, Fig. 10.17 can be a very useful tool for preliminary design.

Remember, too, that zeros can have a considerable effect on system response, and that a root locus gives no information about closed-loop zero location. Computer simulation is therefore generally used to check the theoretical performance of a proposed design, as was noted earlier in the context of continuous systems.

Output

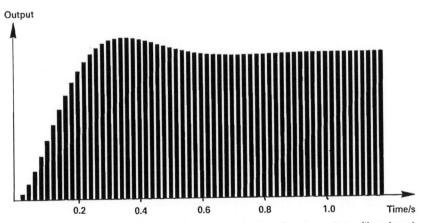

Fig. 10.18 Simulated step response of a second-order discrete system with poles at $c = 0.89$ and $\theta = \pm10°$.

Summary

The pulse transfer function $G(z)$ of a zero-order hold, a continuous process $G_p(s)$ and a sampler at its output is given by the expression:

$$G(z) = \frac{(z-1)}{z} Z\left[\frac{G_p(s)}{s}\right]$$

The z-plane pole positions of $G(z)$ are related to the s-plane pole positions o $G_p(s)$ by the expression $z = e^{sT}$. There is no such generally valid relationshi $\mid$ between the zeros.

Sampled-data pulse transfer functions are generally evaluated by compute software, or from tabulated results such as Table 10.1. To derive such a puls $\mid$ transfer function from first principles, the following procedure is followed:

(1) Form the Laplace transform of the step response $G_p(s)/s$ of the plant.
(2) Evaluate, or look up in standard tables, the z-transform of the sampled ver sion of this continuous step response. This z-transform is denoted $Z[G_p(s)/s]$.
(3) The pulse transfer function of hold, plant and output sampler is then give $\mid$ by the expression

$$G(z) = \frac{(z-1)}{z} Z\left[\frac{G_p(s)}{s}\right]$$

For the configuration of Fig. 10.7, the closed-loop transfer function is given by th $\mid$ expression

$$\frac{KG(z)}{1 + KGH(z)}$$

where

$$GH(z) = \frac{(z-1)}{z} Z\left[\frac{G_p(s)\,H(s)}{s}\right]$$

The closed-loop expression is therefore of the familiar form: 'forward path ove $\mid$ one plus open loop'.

The normal rules of root-locus construction can be used to determine, from th $\mid$ open-loop pole–zero configuration, the way in which z-plane closed-loop pol $\mid$ positions vary as the gain K is increased from a low value. As was the case in th $\mid$ s-plane, the root-locus technique gives no information about closed-loop zeros.

Once a closed-loop pulse transfer function is known, the tabular method o $\mid$ Chapter 9 (or a computer simulation based upon it) can be used to simulat $\mid$ system response to an arbitrary input.

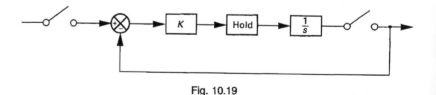

Fig. 10.19

Problems

10.1 In a control system similar to Fig. 10.8 the amplifier and motor combined have a transfer function $G_p(s) = 4/(2 + 3s)$ and the controller gain $K = 3$. What is:

(a) the open-loop pulse transfer function at a sampling frequency of 2 Hz?

(b) the closed-loop pulse transfer function at a sampling frequency of 10 Hz?

10.2 The forward path transfer function of a unity feedback digital control system is $0.8/(z - 0.9)$. What are the steady-state output y_{ss} and the steady-state error e_{ss}?

10.3 Show that the steady-state (low-frequency) gain of a digital compensator $(z + a)/(z + b)$ is equal to $(1 + a)/(1 + b)$.

10.4 Tabulate and sketch the closed-loop step response of the system of Fig. 10.19 for $K = 1$ and a sampling interval of 0.5 s.

10.5 The controller gain K of the system of Fig. 10.12 (Worked Example 10.3) is set to 0.5. Use Fig. 10.17 to estimate the peak overshoot in the discrete step response. Tabulate and sketch the discrete unit step response of the system to check your estimate. What is the nature of the loop output *between* sampling instants in this example?

10.6 A plant modelled as

$$G_p(s) = \frac{20}{s(s + 10)}$$

is combined with a hold and sampler in a digital control system using a sampling frequency of 50 Hz. Find, using Table 10.1, the z-plane pole and zero locations of the corresponding pulse transfer function.

11 Digital control III. Introduction to digital design

Objectives
☐ To apply the techniques developed in Chapters 9 and 10 to the design of simple, single-loop, digital control systems.
☐ To present the digital three-term controller.
☐ To explain how a discrete controller or compensator can be derived from a continuous prototype design.
☐ To introduce the frequency domain approach to discrete controller design.
☐ To describe some approaches to direct discrete design.

The general model of the previous chapter (where two sampled signals were compared to form the error signal) and the new version of Fig. 11.1(a) (where the error signal is formed from two continuous signals and *then* sampled) are equivalent for our purposes. The important thing to remember is that discrete models relate input and output samples only. Behaviour between sampling instants (for those parts of the system where the notion is meaningful) must be considered separately.

Figure 11.1(a) shows a unity feedback, single-loop digital control system. The controller is represented by the discrete transfer function $D(z)$, the DAC is modelled as a zero-order hold, and the plant is represented by the continuous transfer function $G_p(s)$. To simplify the diagram, only one sampler is shown, operating on the error signal. This convention is the norm, and is functionally equivalent to the diagrams of the previous chapter, where samplers were included at the input, the output, and in the feedback path.

Figure 11.1(b) shows the purely discrete version of the system block diagram, using the pulse transfer function $G(z)$ to model the discrete-time behaviour of the hold plus plant. All signals in this model are sample sequences, and the closed-loop transfer function relating input and output samples is

$$\frac{D(z)\,G(z)}{1 + D(z)\,G(z)}$$

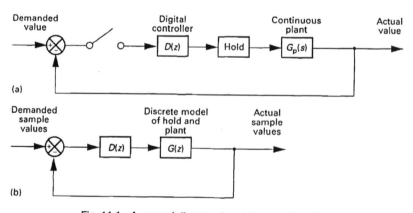

Fig. 11.1   A general discrete closed-loop configuration.

This chapter will cover briefly three different approaches to the general problem of deriving a controller transfer function $D(z)$ to force a given plant to conform to a particular design specification:

A book this size cannot attempt to cover all the techniques in common use. General principles will be emphasized here, leaving the fine details of particular approaches to more specialized and more comprehensive texts. In practice, the choice of one particular technique rather than another is often heavily influenced by personal preference or historical precedence.

(a) the conversion of a continuous compensator or controller design $D(s)$ into an approximately equivalent discrete form $D(z)$;
(b) the use of frequency-response methods to determine an appropriate form of $D(z)$;
(c) the direct discrete design of $D(z)$ based on a knowledge of the pulse transfer function $G(z)$ of the plant.

Each of these approaches is appropriate in particular circumstances, and each is widely used in practice.

The digital PID controller

As a first example of converting a tried and trusted continuous design to discrete form, consider the three-term controller. The differential equation modelling a PID controller was given in Chapter 8 as

$$u(t) = K\left[e(t) + \frac{1}{T_i}\int e(t)\,dt + T_d\frac{de}{dt}\right]$$

where $u(t)$ is the controller output and $e(t)$ is the error signal. Using the discrete approximations to integration and differentiation derived in Chapter 9, the transfer function of a digital PID controller can be written immediately as

$$D(z) = \frac{U(z)}{E(z)} = K\left[1 + \frac{Tz}{T_i(z-1)} + \frac{T_d(z-1)}{Tz}\right]$$

or

The term 'digital PID controller' conforms to conventional usage. Strictly speaking, all the models of this chapter are discrete. They become digital (as defined in Chapter 9) when implemented using digital technology of one form or another – a process which, of necessity, involves both the quantization of measurements and the numerical rounding of values used in the computer algorithm(s).

$$D(z) = K\frac{z^2\left(1 + \frac{T_d}{T} + \frac{T}{T_i}\right) - z\left(1 + \frac{2T_d}{T}\right) + \frac{T_d}{T}}{z(z-1)}$$

A common approach to controller tuning is to set $T_i = 4T_d$, as was the case with the continuous version. Then the digital controller has a double zero located at $z = 2T_d/(2T_d + T)$, where T is the sampling period. For process control using a PID controller, T_i and T_d are often set according to the Ziegler and Nichols tuning rules given in Chapter 8. Typical rules of thumb for selecting the sampling period are that it should be less than a third of the dead time (when using the reaction curve method) or between one-eighth and one-sixteenth of the ultimate period (when using the continuous cycling method).

Do not confuse the use of T in Chapter 8 to represent the process dead time with the use of T in Chapter 9 onwards to represent the sampling interval.

From the previous equation, the transfer function of the PID controller can be written

$$D(z) = \frac{U(z)}{E(z)} = \frac{a_0z^2 - a_1z + a_2}{z(z-1)}$$

where the values of the numerator coefficients can be determined from the controller parameters.

Converting this to a difference equation in the usual way gives

$$u_n = u_{n-1} + a_0 e_n - a_1 e_{n-1} + a_2 e_{n-2}$$

This equation may be implemented directly as a computer algorithm to derive successive output values u_n which are used to position an actuator, such as a valve, directly. The output of the controller is directly proportional to the absolute position of the actuator, and the algorithm is known as an *absolute* or *positional* PID.

For many actuators – a stepper motor, for example – it is more convenient if the actuating signal is proportional to the incremental change required in actuator position. If the incremental controller output is written $y_n = u_n - u_{n-1}$, we have the *incremental PID* algorithm.

$$y_n = a_0 e_n - a_1 e_{n-1} + a_2 e_{n-2}$$

The PID algorithms just presented are widely used in practice. Others exist in which slightly different approximations to integration or differentiation are used.

It is often important that control should be transferred from automatic to manual mode without any sudden discontinuity in controller output. One advantage of the incremental algorithm is that it automatically ensures such 'bumpless transfer' since the existing value of controller output is used as the initial value for the new mode. (Bumpless transfer can, of course, be programmed into a positional PID controller.)

Worked Example 11.1 An alternative, discrete approach to integration, this time using trapezoidal rather than rectangular strips to approximate the area under a curve, is illustrated in Fig. 11.2. Write down the difference equation relating integrator output $y[n]$ and input $e[n]$, and derive the transfer function in z of the trapezoidal integrator.

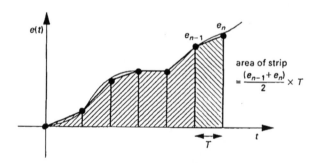

Fig. 11.2 The trapezoidal approximation to integration.

Solution The area of the trapezoidal strip between the $(n-1)$th and nth samples is

$$\frac{(e_n + e_{n-1})T}{2}$$

At each sampling instant the new integrator output y_n is calculated by adding the area of this strip to the previous output y_{n-1}. Hence we can write

$$y_n = y_{n-1} + \frac{(e_n + e_{n-1})T}{2}$$

Converting to sequence notation gives

$$y[n] = y[n-1] + \frac{(e[n] + e[n-1])T}{2}$$

Transforming, we have

$$Y(z) - z^{-1}Y(z) = \frac{[E(z) + z^{-1}E(z)]T}{2}$$

or

$$\frac{Y(z)}{E(z)} = \frac{T(1 + z^{-1})}{2(1 - z^{-1})}$$

Multiplying numerator and denominator by z gives the more usual form of the transfer function for the trapezoidal approximation to integration as

$$G_i(z) = \frac{T}{2} \frac{(z+1)}{(z-1)}$$

Note that the trapezoidal integrator, like the rectangular approximation derived in Chapter 9 (and the pulse transfer function of a continuous integrator) has a pole at $z = 1$. Problem 9.6 was yet another illustration of this association between integrating action and a pole at $z = 1$. We shall return to this important point later.

Digitizing other continuous designs

The above approach to a three-term digital controller is an instance of a more general technique, in which a continuous controller or compensator transfer function is converted to a discrete approximation by directly substituting an appropriate expression in z for the variable s. The procedure corresponds to converting the compensator transfer function $D(s)$ into a differential equation, using an appropriate discrete approximation to differentiation each time it is required, and converting in turn to a discrete transfer function $D(z)$. The discrete compensator $D(z)$ can then be implemented by computer, the control action generated by the algorithm being applied to the plant through a hold in the normal way.

The commonest substitutions used are

$$s \rightarrow \frac{(z-1)}{Tz}$$

as used (implicitly) in the above discrete PID controller, and

$$s \rightarrow \frac{2}{T} \frac{(z-1)}{(z+1)}$$

derived by inverting the transfer function

$$\frac{T}{2} \frac{(z+1)}{(z-1)}$$

of the trapezoidal discrete integrator. (I shall refer to these two substitutions as the 'rectangular' and 'trapezoidal' rules, after their respective approaches to discrete integration.)

So, for example, the lead compensator designed for the ship steering system in Chapter 8

Since

$$\frac{T(z+1)}{2(z-1)}$$

is an approximation to ideal integration represented by the Laplace transfer function $1/s$, the reciprocal expression

$$\frac{2(z-1)}{T(z+1)}$$

is an approximation to ideal differentiation represented by the Laplace transfer function s.

$$D(s) = \frac{1 + s/0.17}{1 + s/1.2} = \frac{5.88s - 1}{0.833s + 1}$$

becomes, using the rectangular rule

$$D_{\text{rect}}(z) = \frac{5.88\left(\dfrac{z - 1}{Tz}\right) + 1}{0.833\left(\dfrac{z - 1}{Tz}\right) + 1}$$

and using the trapezoidal rule

$$D_{\text{trap}}(z) = \frac{5.88\,\dfrac{2}{T}\dfrac{(z - 1)}{(z + 1)} + 1}{0.833\,\dfrac{2}{T}\dfrac{(z - 1)}{(z + 1)} + 1}$$

For comparatively high sampling rates – around 40 to 50 times the closed-loop 3 dB bandwidth, say, as discussed in Chapter 9 – the behaviour of the complete closed-loop system using either of these discrete controllers will differ only slightly from the behaviour of the original autopilot system. As the sampling rate is reduced, however, differences in behaviour between the discrete designs and the original system – as well as between the two different discrete designs – become more and more pronounced. Such protoype designs are most easily compared and evaluated using computer tools, as discussed in Chapter 12. Figure 11.3 shows the results of a computer simulation of the closed-loop step response of each discretized design, together with that of the continuous design. The sampling rate for the discrete simulations is 1 Hz – about six times the closed-loop 3 dB bandwidth, and therefore near the lower limit of practical sampling rates as discussed in Chapter 9. Substituting $T = 1$, the compensator transfer functions become approximately:

A continuous lead compensator adds a real pole and zero in the s-plane, with the zero closer to the origin, $s = 0$. Similarly, a discrete lead compensator adds a real pole–zero pair in the z-plane, with the zero closer to the point $z = 1$. A lag compensator has the relative positions of the pole and zero interchanged in both cases.

$$D_{\text{rect}}(z) = \frac{3.75z - 3.20}{z - 0.45}$$

(a pole at $z = 0.45$ and a zero at $z = 0.85$)

and

$$D_{\text{trap}}(z) = \frac{4.8z - 4.04}{z - 0.25}$$

(a pole at $z = 0.25$ and a zero at $z = 0.84$)

Note that using the trapezoidal rule for discretization gives a response closer to that of the continuous system than using the rectangular rule. This will be the case in general, since the trapezoidal rule gives a closer approximation to the ideal mathematical operations of integration and differentiation than the rectangular rule.

This can be seen most easily by comparing Figs 9.11 and 11.2 for rectangular and trapezoidal integration. The latter is clearly a closer approximation to the area under the curve.

Calculating responses like those of Fig. 11.3 by hand would be very time-consuming, involving the following steps:

(1) Decompose the third-order transfer function representing ship dynamics into

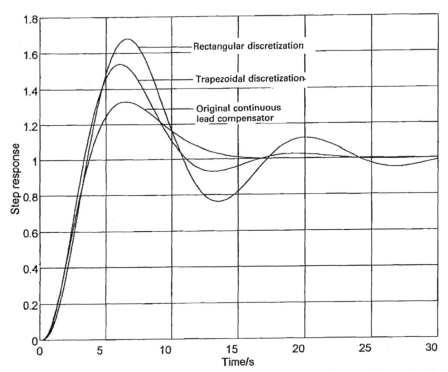

Fig. 11.3 Step responses of the ship autopilot for (a) the continuous design of Chapter 7, (b) a discretized version using the trapezoidal rule, and (c) discretization using the rectangular rule.

a sum of partial fractions as described in Chapter 5.

(2) Convert each term into an appropriate pulse transfer function using the process described in Chapter 10. Recombine into a rational plant pulse transfer function of polynomials in z.

(3) Calculate the closed-loop pulse transfer function with the digital compensator included.

(4) Convert the closed-loop transfer function into a difference equation as described in Chapter 9.

(5) Use the difference equation to simulate the step response as described in Chapter 10.

Discretization using the substitutions

$$s \rightarrow \frac{(z-1)}{Tz} \quad \text{or} \quad s \rightarrow \frac{2}{T}\frac{(z-1)}{(z+1)}$$

is perhaps commonest when sampling rates are kept fairly high, so that digital system behaviour does not diverge too much from that of the original continuous design. Even so, however, certain fundamental differences between discrete and continuous operations must always be kept in mind, as illustrated in the following example.

Worked Example 11.2 What would be the transfer function of a discrete P + D controller using the trapezoidal rule? What problem would arise with the control action generated if this transfer function were implemented?

Solution Substituting for

$$s \to \frac{2}{T} \frac{(z - 1)}{(z + 1)}$$

in the P + D transfer function

$$K(1 + T_d s)$$

gives a corresponding discrete version

$$K\left[1 + \frac{2T_d}{T} \frac{(z - 1)}{(z + 1)}\right] = K\left[\frac{z(T + 2T_d) + (T - 2T_d)}{T(z + 1)}\right]$$

This transfer function has a pole at $z = -1$, which is on the stability borderline, and corresponds therefore to a sequence of samples alternating in sign but with constant magnitude. A P + D controller with a non-decaying component like this in its response would be highly undesirable! For this reason, the substitution $s \to 2(z - 1)/T(z + 1)$ must not be used when the continuous design involves a free differentiator. No such problem arises when the order of the numerator of the continuous transfer function is equal to, or less than, the order of the denominator – a statement equivalent to the condition that there is no free differentiator term. (In fact, many practical implementations of derivative action involve 'smoothed' differentiation fulfilling the latter condition – as described in Chapter 8 and illustrated by Problem 9.5.)

Digital controllers or compensators with (stable) poles *near z* = −1 are also undesirable, since although the oscillatory transient response components die away, they can result in excessive control action or actuator wear. This phenomenon will be taken up again later in the context of direct digital design.

Zero-order hold equivalent discretization

An alternative to substituting directly for s in the compensator transfer function $D(s)$ is to find a transfer function $D(z)$ whose step response is identical to samples of the step response of $D(s)$. This is equivalent to using as $D(z)$ the pulse transfer function of $D(s)$ combined with a hold at the input and a sampler at the output as illustrated in Fig. 11.4. That is,

$$D(z) = \frac{(z - 1)}{z} Z\left[\frac{D(s)}{s}\right]$$

The technique is therefore known as *zero-order hold equivalent* or *step response invariant* design. As before, the behaviour of a control system using a compensator discretized in this way approaches that of the original design for high sampling rates, but may deviate considerably at lower sampling rates. Because manual computation of pulse transfer functions is so laborious, computer tools would again normally be used to determine $D(z)$ from a prototype $D(s)$. In the case of the ship autopilot, the discrete compensator with $T = 1$ using this method is approximately

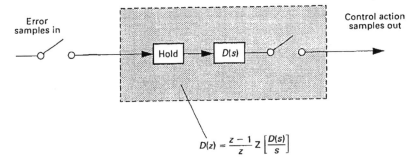

$$D(z) = \frac{z-1}{z} Z\left[\frac{D(s)}{s}\right]$$

Fig. 11.4 Zero-order hold equivalent discretization.

$$D(z) = \frac{7z - 6.3}{z - 0.3}$$

Not surprisingly, there is yet again a real pole and zero, with the zero closer to the point $z = 1$. The precise locations (pole at $z = 0.3$, zero at 0.9) differ slightly from those obtained with the two previous methods of discretization of $D(s)$.

Discretization in the frequency domain

A rather different way of discretizing a continuous design is to move into the frequency domain. This design method is based on an important property of discrete linear systems:

> In the steady state, a discrete linear system will process an input sampled sinusoid so as to produce an output sampled sinusoid at the same frequency, but differing, in general, in amplitude and phase.

This is the discrete equivalent of the frequency preservation property of linear systems.

A discrete linear system therefore possesses a frequency response function which specifies amplitude ratio and phase shift for all frequencies. (Because sample sequences taken from sinusoids at multiples of the sampling rate are indistinguishable, the frequency response of a discrete linear system is periodic, repeating at multiples of the sampling rate as described in Appendix 3.)

The frequency response of a discrete linear system can be derived immediately from its transfer function in z. Recall that multiplication by z^{-1} corresponds to a time delay of one sampling interval, T. Now, if a sequence consists of samples of a sinusoid at frequency ω, represented by a phasor $\mathbf{X}$, then the sequence delayed by T seconds corresponds to samples of the delayed sinusoid $\mathbf{X}e^{-j\omega T}$. Similarly, multiplication of a sinusoidal sequence by z corresponds to operating on the underlying sinusoid with the phasor operator $e^{j\omega T}$. So simply replacing z by $e^{j\omega T}$ converts a transfer function in z to a frequency response function relating steady-state amplitude ratio and phase shift of input and output sinusoidal sequences.

The term $e^{-j\omega T}$ is the phasor delay operator from Chapter 2. The conversion of a transfer function in z into a frequency response function is discussed in more detail in Appendix 3. See also Bissell, C.C., 'A phasor approach to digital filters', Int. J. Elec. Eng. Ed., Vol. 24, pp. 197–213 (July 1987).

Incidentally, this result provides an easy way of calculating the steady-state (low-frequency) gain of an arbitrary discrete linear system. Steady-state gain is obtained by substituting $\omega = 0$ into the frequency transfer function – that is, by substituting $z = e^{j0} = 1$ directly into the transfer function in z.

Hence the result required in Problem 10.3 follows directly.

Worked Example 11.3   Derive an expression for, and sketch, the amplitude ratio of the two-term moving averager, introduced in Chapter 9, with the transfer function

$$G(z) = \frac{(1 + z^{-1})}{2}$$

Show that the amplitude response repeats at intervals equal to the sampling frequency, and that over the range from zero to half the sampling frequency the averager acts as a low-pass filter.

Solution   The frequency response function is obtained by replacing z by $e^{j\omega T}$

$$G(j\omega) = \frac{(1 + e^{-j\omega T})}{2} = e^{-j\omega T/2}\frac{(e^{j\omega T/2} + e^{-j\omega T/2})}{2}$$

The second part of this expression can now be rewritten in trigonometrical form, using the identity

This identity follows immediately from Euler's classic result: exp($j\theta$) = cos θ + j sin θ.

$$\cos\theta = \frac{(e^{j\theta} + e^{-j\theta})}{2}$$

Hence we can write

$$G(j\omega) = e^{-j\omega T/2}\cos(\omega T/2)$$

and the amplitude ratio is given by the expression

$$|G(j\omega)| = |e^{-j\omega T/2}| \times |\cos(\omega T/2)|$$

Now, the modulus of the first term is unity, so the amplitude ratio of the averager is simply

$$|G(j\omega)| = |\cos(\omega T/2)|$$

This is sketched in Fig. 11.5, from which it can be seen that (i) the response repeats at multiples of the sampling frequency, and (ii) over the range zero to half the sampling frequency, the averager behaves like a low-pass filter.

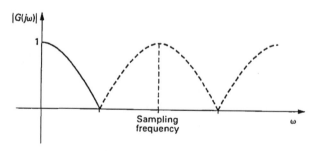

Fig. 11.5   Frequency response of the two-term averager.

Discrete compensators may be designed in the frequency domain in a way very similar to continuous compensators. The crucial point to remember is that the *discrete* frequency response $G(z)|_{z \to e^{j\omega T}}$ of the plant pulse transfer function often differs significantly from the *continuous* plant frequency response $G_p(s)|_{s \to j\omega}$ over the frequency range of interest, as a result of the presence of the zero-order hold. One way of looking at this difference is as follows. The zero-order hold can be viewed as introducing an additional delay of $T/2$ on average, as shown in Fig. 11.6. This delay can be modelled as an additional phase lag, which will tend to reduce stability margins and increase step response overshoot. In the case of the ship autopilot, the expected reduction to the phase margin due to the hold (at a crossover frequency of about 0.5 rad) would be around $\omega T/2 = 0.25$ rad or about 14° at a sampling interval of 1 s. So even a discrete compensator with a frequency response *identical* to that of the original $D(s)$ would result in a phase margin reduced to around 30°. In general, a discrete compensator $D(z)$ may need a quite different frequency response from that of a continuous prototype if the same stability margins are to be achieved with a given plant. And as might be expected, the lower the sampling rate, the greater the difference between the discrete and continuous cases.

This is a rather rough-and-ready approach, but can be very useful in giving a first estimate of the effects of digitizing a continuous design.

The frequency domain approach to digital control system design is particularly common when using the 'trapezoidal' substitution

$$s \to \frac{2}{T} \frac{(z - 1)}{(z + 1)}$$

This is the result of a special relationship between the frequency response of a continuous compensator and its discrete counterpart, when using this method. Suppose that a compensator $D(s)$ is converted into a discrete approximation $D(z)$ using this substitution. The discrete frequency response function is given by substituting $e^{j\omega T}$ for z – in other words, by substituting

This substitution is often known as Tustin's rule, after the British control engineer Arnold Tustin who proposed it in 1947. See: Bissell, C.C., 'Pioneers of control. An interview with Arnold Tustin', *IEE Review*, June 1992, pp. 223–226.

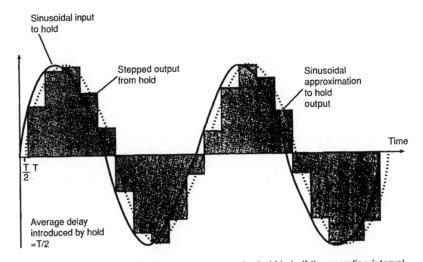

Fig. 11.6 Average delay introduced by a zero-order hold is half the sampling interval.

$$s \rightarrow \frac{2}{T} \frac{(e^{j\omega T} - 1)}{(e^{j\omega T} + 1)}$$

directly in $D(s)$.

The right-hand side of this expression can be converted into a much more convenient form by means of Euler's formula, $e^{j\theta} = \cos\theta + j\sin\theta$. After some manipulation we can write:

Problem 11.3 asks for the derivation of this expression.

$$s \rightarrow \frac{j2}{T} \tan\left(\frac{\omega T}{2}\right)$$

Suppose now that the discrete compensator is to introduce a defined amplitude ratio and phase shift at frequency ω_d. The *discrete* frequency response at ω_d is obtained by substituting

$$s \rightarrow \frac{j2}{T} \tan\left(\frac{\omega_d T}{2}\right)$$

Inserting numerical values gives a simple demonstration of the effect, but we can also proceed analytically. Substituting for the 'angular sampling frequency' $\omega_s = 2\pi/T$ we can write

$$\frac{\omega_c}{\omega_s} = \frac{1}{\pi} \tan \frac{\pi\omega_d}{\omega_s}$$

If $\omega_d \ll \omega_s$, then

$$\frac{1}{\pi} \tan \frac{\pi\omega_d}{\omega_s} \approx \frac{\omega_d}{\omega_s}$$

and $\omega_c \approx \omega_d$. So for sufficiently high sampling rates the distortion of the frequency scale can be ignored.

into $D(s)$. Now, recall that the substitution $s \rightarrow j\omega_c$ gives the frequency response of the *continuous* controller at ω_c. By comparing the forms of these two substitutions we obtain the following result for discretization by Tustin's rule.

The frequency response of a continuous compensator $D(s)$ at a frequency ω_c is identical to that of the discrete version

$$D(z) = D(s)\Big|_{s \rightarrow \frac{2}{T}\frac{(z-1)}{(z+1)}}$$

at a frequency ω_d where

$$\omega_c = \frac{2}{T} \tan\left(\frac{\omega_d T}{2}\right)$$

From this expression it follows that for frequencies well below the sampling rate, $\omega_c \approx \omega_d$. For example, suppose that the sampling rate is 10 Hz, so that $T = 0.1$, and that the digital compensator should ultimately have a specified response at 1 Hz ($2\pi \, \text{rad s}^{-1}$). Substituting in the expression, we have

$$\omega_c = 20\tan(0.1\pi) = 6.5 \, \text{rad s}^{-1} = 1.03 \, \text{Hz}$$

In other words, at a tenth of the sampling frequency there is little difference between the frequency response of the discrete and continuous compensators.

For frequencies up to about a tenth of the sampling rate, then, applying Tustin's rule to a continuous compensator results in a digital equivalent with a frequency response very similar to that of the original compensator. This is illustrated in Fig. 11.7, which shows the relationship between the continuous and discrete frequency scales up to around one-third of the sampling rate.

To reiterate an earlier point, this does *not* mean that an existing continuous compensator can be replaced by a digital version with a very similar (discrete) frequency response without a significant effect on closed-loop behaviour. The frequency response corresponding to the pulse transfer function $G(z)$ can differ markedly from that of the plant $G_p(s)$ alone (because $G(z)$ models the presence of the zero-order hold, and hence the application to the plant of a stepped waveform rather than a smoothly varying one). For example, with a sampling rate

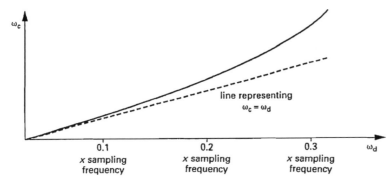

Fig. 11.7 Distortion of the frequency scale introduced by the substitution $s \to 2(z - 1)/T(z + 1)$.

of 1 Hz – that is, more than six times the closed-loop 3 dB bandwidth of around 1 rad s^{-1} – the frequency response of the ship autopilot compensator digitized by Tustin's rule does not differ greatly from its continuous counterpart over the range of frequencies of interest. But as already seen in Fig. 11.3, the simulations of the continuous and discrete designs show considerable differences in step-response overshoot.

To achieve given stability margins with a discrete compensator, the prototype continuous compensator parameters must therefore be determined on the basis of the discrete frequency response of the plant *pulse transfer function, G(z)* – and *not the original continuous plant frequency response $G_p(j\omega)$*. This can done with the aid of a computer design package in the following way:

(1) Compute the discrete frequency response $G(z)|_{z \to e^{j\omega T}}$ of the hold, plant and sampler for the chosen sampling frequency over the frequency range of interest. Display the result in your preferred way (as a Nichols, Bode or Nyquist plot).
(2) From the results of (1), design a continuous compensator $D(s)$ to satisfy the specified stability margins, using standard techniques such as the one outlined in Chapter 8.
(3) Convert $D(s)$ into $D(z)$ using Tustin's rule, perhaps using a built-in computer function. Check that the loop still satisfies stability margins, and implement digitally.

This procedure implies that the sampling rate is high enough for the difference in the frequency responses of $D(s)$ and $D(z)$ to be ignored over the frequency range of interest. A further refinement is necessary if sampling rates have to be kept so low that the distortion of the frequency scale of Fig. 11.6 is significant. In such cases, allowance is made at stage 2 for the subsequent effect on the frequency response of the substitution

$$s \to \frac{2}{T} \frac{(z - 1)}{(z + 1)}$$

This is known as *pre-warping* the design of $D(s)$. In other words, the formula

231

$$\omega_c = \frac{2}{T}\tan\left(\frac{\omega_d T}{2}\right)$$

introduced earlier is used to determine the critical prototype frequencies of $D(s)$ which, *after* discretization, will give exactly the characteristics of $D(z)$ required to satisfy the specification. In many cases, however, sampling rates will be high enough for pre-warping to make little difference to final performance.

<table>
<tr><td>**Worked Example 11.4**</td><td>Suppose that a digital compensator has to introduce a specified phase shift at a critical frequency of 2 Hz, in a system using a sampling frequency of 10 Hz. What should be the corresponding critical frequency of the continuous design, before discretization using Tustin's rule?

Solution   Substituting appropriate values we have

$$\omega_c = 20\tan(0.2\pi) = 14.5\,\text{rad s}^{-1} = 2.3\,\text{Hz}$$

So if the compensator $D(s)$ is designed to give a particular phase shift at a frequency of 2.3 Hz, then $D(z)$ will give precisely this value (as well as the same amplitude ratio as the continuous compensator) at the desired frequency of 2 Hz.</td></tr>
</table>

All the methods considered so far in this chapter rely on converting an existing, or prototype, continuous design into a discrete version. Whatever method of discretization is adopted, it is vital to remember that the performance of the resulting digital system may differ substantially from that of the prototype unless high sampling rates are employed. Thorough simulation, testing, and iterative refinement of the design are therefore of utmost importance – with computer simulation playing an indispensable role.

Quantization and numerical rounding effects are, of course, just as important in direct discrete design, the topic of the next section. Quantitative consideration of these effects is beyond the scope of this book: for a reasonably accessible treatment see Franklin, Powell and Workman (1990).

Other factors which may need to be taken into consideration are quantization and numerical rounding – in other words, the consequences of the finite wordlength of the *digital* implementation of the *discrete* design. Quantization in data conversion can be treated as introducing a particular type of noise into the system, and can be modelled explicitly as a noise source. The use of a finite wordlength when implementing controller algorithms means that it may not be possible to realize the precise design values of controller coefficients. Yet another iteration of simulation may therefore be needed, with the rounded realizable coefficient values substituted for the ideal ones. Both constraints become particularly significant if wordlengths need to be kept low for cost reasons in mass production items using cheap microprocessors and data converters. As is always the case in engineering, application of theory must be tempered with a thorough appreciation of those practical constraints which have not been incorporated into the original system model.

Direct discrete design

A controller or compensator transfer function $D(z)$ may be derived directly from a knowledge of the plant pulse transfer function $G(z)$ without any reference whatsoever to a continuous prototype. Recall from Chapter 8 that a control system specification is likely to be couched in such terms as steady-state error; maximum step response overshoot and settling time; and error when tracking an input changing at a constant rate. In a continuous design, the steady-state aspects would be satisfied by ensuring sufficiently high controller gain, and/or sufficient integrators in the forward path; while the required closed-loop dynamic behaviour might be translated into the need for dominant s-plane poles within a particular region of the s-plane.

This approach can be applied directly to discrete systems. It has already been noted that all the discrete approximations to integration possess a pole at $z = 1$. This $(z - 1)$ term in the transfer function denominator can be interpreted in the following way. Suppose that the input is constant. Then, after any decaying transient terms have died away, the nth output sample is formed by simply adding the input (scaled by a gain factor if appropriate) to the $(n - 1)$th output sample. The output thus accumulates steadily in value as long as the input is present. In other words, transfer functions with a $(z - 1)$ denominator term possess precisely the integrating function needed to eliminate steady-state error, as described in Chapter 7. Similarly, a $1/(z - 1)^2$ term is necessary in the forward path of a discrete system if tracking is to be achieved without steady-state error.

One approach to direct digital control system design is therefore to include sufficient poles in the forward path at $z = 1$ to satisfy steady-state requirements, and to add zeros at appropriate points in the z-plane to 'attract' the root locus to those regions which imply satisfactory dynamic response. As with continuous systems, this approach is feasible only with computer software to plot the loci and to investigate the effects of parameter modifications.

An alternative approach is more analytical. Assuming a unity feedback configuration as was the case in Fig. 11.1, and writing the closed-loop pulse transfer function as $F(z)$, we have

$$F(z) = \frac{D(z)G(z)}{1 + D(z)G(z)}$$

Hence

$$D(z) = \frac{F(z)}{G(z)[1 - F(z)]}$$

This equation can be used as a basis for deriving an appropriate $D(z)$, given a desired $F(z)$. The important point is to choose a closed-loop transfer function $F(z)$ which is realizable from both theoretical and practical points of view. Limitations of space mean that full details of the many approaches to direct digital design cannot be given here. Instead, we conclude by considering a number of important factors which will *always* need to be taken into account.

To eliminate steady-state error in response to a step input, an integrator must be present in the forward path – 'upstream' of disturbances if it is to eliminate the effects of these too. Elimination of error under steady-state *tracking* conditions requires *two* integrators in the forward path. Refer to Chapter 7 for details.

The apparent power of the direct digital design technique can be seductive. It is easy to manipulate the algebraic expressions to obtain an $F(z)$ of virtually any desired form. But the algebra may hide something quite absurd: a non-causal controller; a requirement for a servomotor which would have to provide far greater torques than is possible for its size; or even an unstable system!

233

Causality constraints

If the plant $G(z)$ introduces a delay of n sample intervals, it is no use choosing an $F(z)$ with a shorter delay -- otherwise the plant would have to respond to a control action before it was applied! This constraint is equivalent to stating that if $G(z)$ has m more poles than zeros, so must $F(z)$.

'Cancellation' issues

A plant might have a pole at, say, $z = 0.95$, and a controller might be designed with a zero at $z = 0.95$ and a new pole at, say, $z = 0.8$. The controller zero 'cancels' the plant pole, and the closed-loop transient response then takes many fewer sampling intervals to reach its steady-state value than would be the case without the controller.

Implicit in many direct digital techniques is the notion that $D(z)$ is chosen so as to 'cancel' the plant dynamics, adding appropriate new poles and zeros to give $F(z)$ the dynamics required.

When applying the direct design formula all the caveats mentioned earlier in Chapters 5, 7 and 8 continue to apply. In practice, 'cancellation' means that a greater control action is applied in order to force the plant to respond more quickly than it otherwise would. Whether or not this is actually possible depends on a whole range of factors which are often not incorporated into the system models. Furthermore, the plant pole is certainly not 'cancelled' as far as *disturbances* downstream of the controller are concerned, so these may still take a long time to die away. Beware, therefore, of proposed designs which apparently 'speed up' response by an excessive amount, particularly where disturbance rejection is important.

Stability issues

This is exactly analogous to the use of velocity feedback or derivative action to stabilize an open-loop unstable system as described in Chapter 6. Details of design approaches to the placement of such zeros are beyond the scope of this book, but can be found in Franklin, Powell and Workman (1990), or Golten and Verwer (1991).

If the plant is open-loop unstable, then merely manipulating the expression

$$D(z) = \frac{F(z)}{G(z)[1 - F(z)]}$$

might well lead to an attempt to cancel a pole outside the unit circle by a zero of $D(z)$. But such cancellation can never be exact, so the closed-loop system would still be unstable. Instead, an appropriate zero must be added elsewhere so that the z-plane root locus passes into the unit circle for an appropriate value of gain.

The ringing pole problem

An important final point to note is a problem which can arise if the plant has a zero between $z = 0$ and $z = -1$. A controller design $D(z)$ in which such a plant zero is cancelled by a pole will have no stability implications, but could still result in completely inappropriate control action being applied to the plant. Poles between $0 < z < -1$ correspond to transient response sequences in which alternate samples have opposite signs (oscillation at half the sampling frequency). A controller with a pole in this region may therefore produce oscillating control action which results in increased actuator wear -- even if the plant output appears to conform to the specification. Such a controller pole lying near $z = -1$ is termed a ringing pole. In the worst case, a ringing controller pole may produce significant 'hidden oscillations' in the plant output. That is, the sampled plant output may conform to specifications, but the behaviour between samples may oscillate widely. A rather artificial example is shown in Fig. 11.8 to illustrate what can happen. Here the plant is modelled as an integrator plus a delay equal to half the sampling interval, and the oscillatory control action gets 'out of step' with the

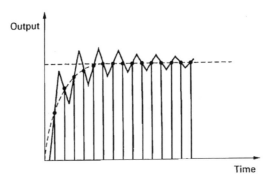

Fig. 11.8 An example of severe inter-sample oscillation undetected in the output samples.

oscillatory plant response. Yet if this system were assessed on its sampled step response only, it might appear to satisfy requirements perfectly!

Direct digital design can be employed successfully provided the basic ideas are applied sensibly. To sum up, always:

Check the final version of $D(z)$ – are there any ringing or unstable poles?

Simulate the system closed-loop response – is the behaviour acceptable, including that between sampling instants?

Consider the magnitude and the nature of the control action – is it practicable with the actuators envisaged?

Simulate the effect of possible disturbances – are the disturbance rejection dynamics acceptable?

Consider the sensitivity of the final design – are any uncertainties in the plant model likely to affect performance significantly?

Much of this advice holds for *any* proposed controller design. But there is, perhaps, more scope with direct digital design for deriving unsuitable forms of $D(z)$ which *appear* to be valid than when using a design based on a continuous prototype.

Design example

Figure 11.9 shows a video tape position control system using a sampling frequency of 100 Hz. A digital compensator of the form

Access to appropriate simulation software is needed in order to work through this example.

$$D(z) = K\frac{(z + a)}{(z + b)}$$

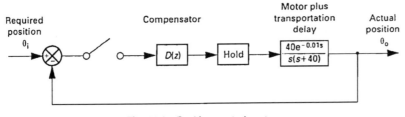

Fig. 11.9 Position control system.

235

is to be designed, where a and b are constants, such that the step response has zero steady-state position error, reaches 1% of its final value after 0.1 s at the most, and exhibits little or no overshoot (maximum of 1%).

Discussion

The pulse transfer function of the hold, plant and sampler is, from computer software or Table 10.1 (incorporating an additional pole at the origin to model the delay of one sampling interval):

$$G(z) = \frac{0.00176(z + 0.80)}{z(z - 1)(z - 0.67)}$$

The specification can be translated into limits on dominant closed-loop pole positions. First, poles must lie within a radius of about 0.6 of the origin: $0.6^{10} \approx 0.006$, so dominant poles in this region imply that the system would settle to within 1% after ten sampling intervals (0.1 s). To satisfy the overshoot constraint, damping ratio must be 0.8 or greater, implying (from Fig. 10.17) that the angle of the dominant poles should be 20° or less.

A naive approach might attempt to cancel the plant pole at $z = 0.67$ with a controller zero, and the plant zero at $z = -0.8$ with a controller pole. The open-loop transfer function would then have poles at $z = 0$ and $z = 1$, and the root locus for varying gain would be identical to that of Fig. 10.13, allowing the closed-loop poles to be positioned in an appropriate region of the z-plane by adjusting the gain. A controller pole at $z = -0.8$, however, would be most unsatisfactory from the ringing pole point of view discussed above, and a compromise has to be found.

The likely effects of various controller pole positions can be investigated using a root-locus package: the plot for a controller transfer function

$$D(z) = K\frac{(z - 0.67)}{(z + 0.3)}$$

If you have access to suitable software, try simulating system performance – particularly control action and inter-sample behaviour – for a controller with a pole cancelling the plant zero at $z = -0.8$.

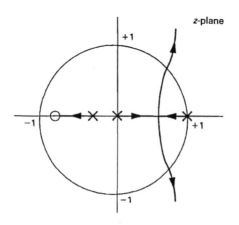

Fig. 11.10 Root-locus plot for the system of Fig. 11.9.

236

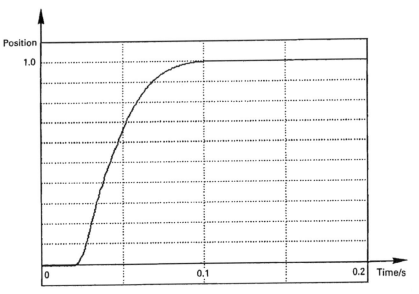

Fig. 11.11 Step response of the proposed design.

The simulation software used to generate these figures assumes a 'realistic' applied step with a finite rise time beginning at $t = 0$. Hence the control action begins after one sampling interval, and not immediately as might be implied by the form of $D(z)$.

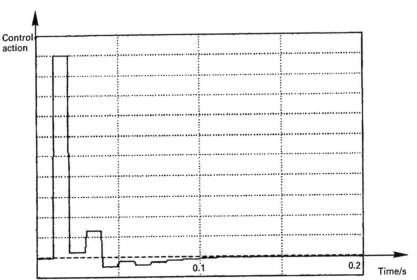

Fig. 11.12 Control action of the proposed design.

is shown in Fig. 11.10. With $K = 100$, the closed-loop poles lie at $z = -0.43, 0.56 \pm j0.20$, and the step response, fulfilling the system specification, is illustrated in Fig. 11.11. Simulated control action is shown in Fig. 11.12, from which the slight oscillatory component can be seen. In practice, such simulation should be followed by a consideration of the practicalities of implementation, prototype testing, and further iteration of controller design if necessary.

Summary

Two basic approaches to the design of a simple, closed-loop digital control system have been introduced. In the first, a continuous controller or compensator design (as a transfer function in s) is converted into a digital version (as a transfer function in z) possessing similar, but not identical, behaviour. Techniques include:

- substituting

$$s \rightarrow \frac{(z-1)}{Tz} \quad \text{or} \quad s \rightarrow \frac{2}{T} \frac{(z-1)}{(z+1)}$$

  into the continuous transfer function $D(s)$;
- using the 'zero-order hold' equivalent

$$D(z) = \frac{(z-1)}{z} Z \left[\frac{D(s)}{s} \right]$$

  as the digital compensator;
- digital frequency response design based on stability margins (with or without pre-warping of critical frequencies).

In each case the behaviour of the digital system may differ considerably from that of the continuous prototype, particularly if the sampling rate is towards the lower end of the practicable range.

In the second approach, the digital controller transfer function $D(z)$ is derived directly from the pulse transfer function of the plant, using root locus or other methods based purely on discrete system models. Because of the discrete nature of the design process, it is particularly important to simulate the final design and to check intersample behaviour and controller output characteristics (control action).

Problems

11.1 The continuous P + D controller of the satellite example of Chapter 6 (Fig. 6.8) is discretized using the rectangular approximation to differentiation described in Chapters 10 and 11. The sampling period is 0.2 s. For $K = 4$, $K_s = 2$ and $T_d = 0.6$ (as in Chapter 6):
   (a) Calculate the closed-loop pulse transfer function.
   (b) Sketch, or simulate using appropriate software, the unit step response of the closed-loop system.
   Compare your sketch with Fig. 6.29 and comment on the differences.

11.2 Show, using appropriate results from Table 10.1, that the plant pulse transfer function of Fig. 11.9 is indeed

$$G(z) = \frac{0.00176(z + 0.80)}{z(z - 1)(z - 0.67)}$$

as stated.

11.3 Derive the expression

$$s \rightarrow \frac{j2}{T}\tan\left(\frac{\omega T}{2}\right)$$

for Tustin's rule (trapezoidal discretization).

11.4 If you have suitable computer software available, simulate the step response of the ship autopilot example using the zero-order hold equivalent compensator

$$D(z) = \frac{7z - 6.3}{z - 0.3}$$

and a sampling rate of 1 Hz. Compare the result with Fig. 11.3.

11.5 A digital compensator is to be designed for the system of Problem 8.3 (Fig. 8.23), using a sampling rate of 50 Hz.
   (a) What would be the discrete compensator $D(z)$ derived by applying Tustin's rule directly to the solution of Problem 8.3?
   (b) Estimate the reduction in uncompensated phase margin resulting from the presence in the discretized version of a zero-order hold.
   (c) Using a frequency domain approach, derive a new $D(z)$ to satisfy the performance specification of Problem 8.3.
   (d) If possible, simulate the system closed-loop step response with the digital compensators of (a) and (c). Comment on the differences.

11.6 If appropriate software is available, investigate a z-plane root-locus approach to designing a digital compensator for the system of Problem 8.3 (Fig. 8.23). K must be at least 20 to satisfy the tracking condition, and the peak closed-loop step response overshoot is limited to 15% as before. (*Hint*: Use a compensator of the form $(z + a)/(z + b)$ to cancel the plant pole at $s = -10$. Take into account the steady-state gain of the compensator.)

11.7 Repeat 11.5(c) using a sampling rate of 100 Hz. If possible, compare the control action in the two cases.

12 Computers and control

Objectives
☐ To introduce a range of computer tools for the design of control systems.
☐ To present some more advanced features of digital controllers and computer-based control algorithms.

In recent years computers have transformed the practice of control engineering. In this final chapter two important changes will be considered briefly: (i) the widespread availability of high-performance computer tools for system modelling and design; and (ii) the commercial exploitation, in modestly priced off-the-shelf controllers, of various 'advanced' techniques formerly limited to highly specialized systems.

Tools for system modelling and design

The design procedures described in this book are essentially the 'classical' single-loop techniques developed in the 1940s and 1950s. All of them – from the frequency response approach of Chapter 4, through the root-locus methods of Chapter 6, to the discrete variants of Chapter 11 – involve an element of trial and error. Designs often have to be refined by the iterative adjustment of controller or compensator parameters such as gain, integral time or time constants. Modern computer tools allow this iteration to be carried out speedily, while retaining the conceptual simplicity of traditional engineering approaches such as Bode plots and Nichols charts. Learning to use such tools with insight and understanding forms an important part of modern engineering education.

Dedicated control system design packages

This chapter is intended to give a flavour of the tools available at the time of writing (1993). For a series of articles giving practical case studies of the use of various commercial packages see *Measurement & Control*, Vol. 25, Nos 5 (June), 7 (September) and 8 (October), 1992.

Many software packages are available which have been developed specifically for control system design. The most sophisticated of these (and, of course, the most expensive) are aimed at specialist designers. They allow the system to be specified graphically, using the mouse, cursor and keyboard to define block diagrams, enter transfer functions, and interconnect components in almost unlimited configurations. Simulation results can be presented in a whole variety of ways: Nichols, Nyquist or Bode plots; s- and z-plane root locus; time response to standard or arbitrary inputs; and so on. Typically, the user will be able to choose from a variety of design techniques, both the classical ones presented here and others beyond the scope of this text. The upper end of the market is represented by packages such as Program CC, Ctrl-C, $MATRIX_x$ or Matlab (the latter two being general engineering simulation packages with special control system design options). More modest control system design packages are also available, at a correspondingly lower cost. Although they restrict the user to the commoner configurations and design techniques, they are perfectly adequate for many pro-

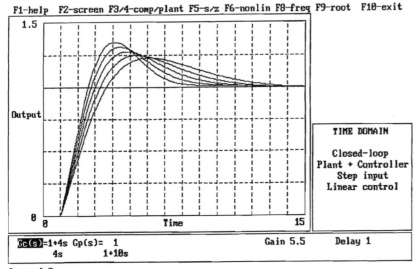

Fig. 12.1 CODAS-II used to simulate the step response of the temperature control
system of Chapter 4 (values of proportional gain between 4 and 6).

fessional control engineering purposes: a widely used example in education and
industry is CODAS-II.

As a first example of the power of such packages, let us go back to the
temperature control example of Chapter 4. A frequency response design like this
example can be carried out quickly and easily using such a tool, with the computer generating the various Bode and Nichols plots needed as part of the design
process. Moreover, the computer package can plot time domain responses without
recourse to second-order approximation. Simulated step responses using CODAS-
II are shown in Fig. 12.1 for the temperature control example with a controller
integral time of 4 s (as derived in Chapter 4) and values of gain K between 4 and 6
(at intervals of 0.5). In Chapter 4, the time domain behaviour was approximated
by a standard second-order system: the simulation shows that the predictions of
this model (25% overshoot for $K \approx 5$, settling to within 2% of final value within
17 s) were reasonably accurate, even though the actual model deviates significantly from second-order behaviour. The great advantage of simulation is that the
effect of slight modifications to gain, integral time and pure time delay can be
investigated with very little effort by the designer.

As a second example of the use of a dedicated control system design package,
Fig. 12.2 shows the printout of a window from Matlab's Simulink package during
the simulation used to generate the step responses of Fig. 11.3. The block
diagrams of the original system, together with the two discretized versions were
built up on screen. A standard Matlab option was used to generate the Tustin
(trapezoidal) discrete transfer function, while computer algebraic manipulation
was used for the simple rectangular substitution. Once constructed, the model can
be used to generate other plots in addition to the step responses of Chapter 11.

Recall that, in the example, a
P + I controller is used to control
a plant modelled by a first-order
lag with a 10 s time constant, plus
a 1 s time delay. Controller and
plant transfer functions, gain and
time delay can be read at the foot
of the screen display.

241

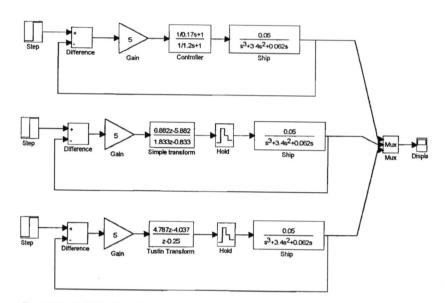

Fig. 12.2   Matlab's Simulink package used to compare the continuous ship autopilot with
rectangular and trapezoidal discretized versions.

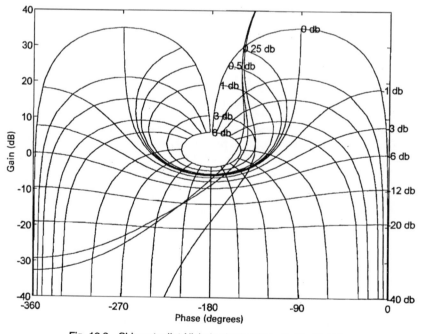

Fig. 12.3   Ship autopilot Nichols charts generated by Matlab.

```
%Calculate the margins
%Find the gain and phase margin for the continuous system
[m,p,w]=bode(Fn,Fd);
[GM,PM,wg,wp]=margin(m,p,w);
GainMargin=20*log10(GM)

GainMargin =

    18.8345

PhaseMargin=PM

PhaseMargin =

    43.6045


%Tustin transformed controller
[m,p,w]=dbode(Fz1n,Fz1d,1);
[GM,PM,wg,wp]=margin(m,p,w);
GainMargin=20*log10(GM)

GainMargin =

    8.6327

PhaseMargin=PM

PhaseMargin =

    31.2583
```

Fig. 12.4 Matlab commands for calculating gain and phase margins.

A Matlab display of the Nichols plots for the three designs is given in Fig. 12.3, while Fig. 12.4 illustrates the Matlab commands for computing the gain and phase margins of the continuous system and the discrete version derived from Tustin's rule. (Note, incidentally, that the Chapter 11 estimate of a phase margin of around 30° for the trapezoidal discretization is confirmed.) Clearly, such plots would be prohibitively time-consuming to plot accurately without the aid of a computer.

Dedicated computer-based tools allow controller design – whether based on frequency response, root locus or other methods – to be carried out interactively, and without the need to carry out laborious calculation, algebraic manipulation, or graph plotting by hand. In addition to the usual closed-loop step or ramp response, other important system variables such as the magnitude of the control action, or the response to disturbances, can be simulated easily. Investigation of the effects of slight changes to system parameters, often prohibitively time-consuming to carry out manually, can be accomplished in a few keystrokes. Needless to say, the user needs to understand the techniques underlying the simulations if a satisfactory design is to be produced!

How well such simulations reflect actual system behaviour will, of course, depend on the validity of the assumptions made in deriving the plant model.

General-purpose mathematics packages

General-purpose mathematical software can also be used for control system design. Into this category come packages like Mathematica, Macsyma and

```
(c3) /* The ship transfer function */
ship:0.05/(s*(s^2+3.4*s+0.062));
```

(d3)
$$\frac{0.050}{s\left(s^2 + 3\,400\,s + 0\,062\right)}$$

```
(c4) /* Set the autopilot gain */
gain:5.0;
```
(d4)   5.000

```
(c5) /* Specify the compensator */
compensator:(1.0+s/0.17)/(1.0+s/1.2);
```

(d5)
$$\frac{5.882\,s + 1.000}{0.833\,s + 1\,000}$$

```
(c6) /* Form the open loop transfer function */
open_loop:combine(expand(gain*compensator*ship));
```

(d6)
$$\frac{1.471\,s + 0.250}{0.833\,s^4 + 3.833\,s^3 + 3.452\,s^2 + 0.062\,s}$$

```
(c7) /* Form the closed loop transfer function from the
         open loop numerator and denominator           */
closed_loop:combine(expand(num(open_loop)/(num(open_loop)+denom(open_loop))));
```

(d7)
$$\frac{1.471\,s + 0.250}{0.833\,s^4 + 3.833\,s^3 + 3.452\,s^2 + 1.533\,s + 0\,250}$$

```
(c8) /* Form the characteristic equation */
characteristic_equation:subst(x,s,denom(closed_loop)=0);
```

(d8) $0.833\,x^4 + 3.833\,x^3 + 3.452\,x^2 + 1.533\,x + 0.250 = 0$

```
(c9) /* Find the roots */
roots:allroots(characteristic_equation);
```
(d9) $[x = -0.310, x = 0.380\,i - 0.355, x = -0.380\,i - 0.355, x = -3.580]$

```
(c10) /* Closed loop transfer function in factored form */
transfer_function:realpart(num(closed_loop)/
(coeff(denom(closed_loop),s,length(roots))*prod(subst(roots[n],s-x),n,1,length(r
C:\maths\macsyma\library1\combin.fas being loaded.
```

(d10)
$$\frac{1.200\,(1.471\,s + 0.250)}{(s + 0.310)\,(s + 3.580)\left((s + 0.355)^2 + 0.144\right)}$$

```
(c11) /* Laplace transform of the step response */
1/s*transfer_function;
```

(d11)
$$\frac{1.200\,(1.471\,s + 0.250)}{s\,(s + 0.310)\,(s + 3.580)\left((s + 0.355)^2 + 0.144\right)}$$

```
(c13) /* and clean up */
step_response:scanmap('float,bfloat(expand(-)));
```

(d13) $-1.545\,e^{-0.355\,t}\sin(0.380\,t) - 2.621\,e^{-0.355\,t}\cos(0\,380\,t) + 1.669\,e^{-0.310\,t} - 0\,049\,e^{-3\,580\,t} + 1\,000$

```
(c14) /* Plot the result */
plot(step_response,t,0,20,"Time/s","Angular position","Ship autopilot step respo
ymin:0.0,ymax:1.5,plotnum:400;
```
(d14)   done
(d15)   done
(c16)

Fig. 12.5   Macsyma commands for calculating autopilot step response by inverse Laplace
transformation.

Mathcad. Such packages can carry out algebraic manipulation, solve differential
equations (numerically or symbolically), and display mathematical results in
various graphical forms. They can thus be used to carry out many of the design
and analysis procedures described in this book. An example is shown in Fig. 12.5,

which is part of a printout of Macsyma being used to compute the analytical form of the ship autopilot step response using the continuous compensator of Chapter 8. The program derives the closed-loop transfer function by carrying out the appropriate algebraic manipulations, then computes the step response by inverse Laplace transformation as described in Chapter 6. Plotting the step response is again a matter of a few keystrokes. Clearly, there are many other applications of such tools to control system design.

Spreadsheets

The previous two types of package are still rather expensive. A useful alternative to those who do not have access to such tools – and an excellent aid to understanding many of the techniques presented in this book – is the spreadsheet. Originally designed for financial modelling, spreadsheets consist essentially of programs to carry out mathematical operations on columns of figures. They are well suited, however, to a number of control applications. For example, Fig. 12.6 shows a spreadsheet computation of the ship autopilot frequency response using the continuous lead compensator design of Chapter 8. Separate columns calculate the amplitude and phase contributions of the individual system components, using built-in arithmetic, trigonometrical and logarithmic spreadsheet functions. The open-loop frequency response is then summed in columns L and M. Rows 7–8 of the spreadsheet (corresponding to loop gain of around 0 dB) show that the phase

	A	B	C	D	E	F	G	H	I	J	K	L	M
1		loop	lead (tau)		lag (tau)		int		2nd (zeta)	nat freq		total	total
2	freq/rad	gain/dB	5.88		0.833				6.83	0.249		total	total
3		12	amp/dB	phase	amp/dB	phase	amp/dB	phase	amp/dB	phase		phase	gain/dB
4													
5	0.2	12	3.7712	49.6	-0.119	-9.46	13.979	-90	-20.8102	-88.15		-138	8.8215
6	0.3	12	6.14021	60.5	-0.263	-14	10.458	-90	-24.3307	-91.57		-135.2	4.004
7	0.4	12	8.1504	67	-0.457	-18.4	7.9588	-90	-26.8487	-94.12		-135.6	0.8033
8	0.5	12	9.84239	71.2	-0.695	-22.6	6.0206	-90	-28.8172	-96.31		-137.7	-1.649
9	0.6	12	11.2862	74.2	-0.968	-26.6	4.437	-90	-30.4397	-98.31		-140.7	-3.685
10	0.7	12										-144.1	-5.46
11	0.8	12										-147.7	-7.055
12	0.9	12										-151.3	-8.519
13	1	12										-154.9	-9.881
14	1.1	12										-158.3	-11.16
15	1.2	12										-161.7	-12.37
16	1.3	12										-164.9	-13.52
17	1.4	12										-168	-14.63
18	1.5	12										-171	-15.68
19	1.6	12										-173.8	-16.7
20	1.7	12										-176.5	-17.67
21	1.8	12										-179.1	-18.62
22	1.9	12										-181.6	-19.53
23	2	12										-184	-20.41
24	2.1	12										-186.2	-21.27
25	2.2	12	22.2019	85.0	-6.595	-61.4	-6.8405	-90	-43.1700	-122.0		-188.4	-22.1
26	2.3	12	22.6458	85.8	-6.694	-62.4	-7.2346	-90	-43.6231	-123.8		-190.4	-22.91
27	2.4	12	23.0135	85.9	-6.987	-63.4	-7.6042	-90	-44.113	-124.9		-192.4	-23.69
28	2.5	12	23.3664	86.1	-7.273	-64.4	-7.9588	-90	-44.5895	-126		-194.3	-24.45

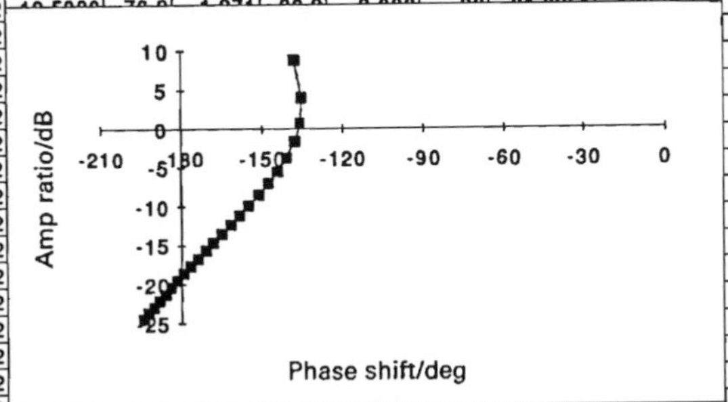

(A plot of Amp ratio/dB versus Phase shift/deg is embedded over rows 10–25.)

Fig. 12.6 Spreadsheet simulation of the ship autopilot.

Formulae for the first- and second-order terms are entered, in standard form, assuming unity low-frequency gain. A first-order lag term hence has amplitude ratio $1/\sqrt{1 + \omega^2\tau^2}$ and a phase shift of $-\arctan(\omega\tau)$; a second-order term has amplitude ratio $\omega_n^2/\sqrt{(\omega_n^2 - \omega^2)^2 + (2\zeta\omega_n\omega)^2}$ and phase shift $-\arctan[2\zeta\omega_n\omega/(\omega_n^2 - \omega^2)]$. The combined loop gain (dB) is entered separately in column B.

	A	B	C	D	E	F	G	H	I
1	input			output	pole positions				
2	x[n]	y[n-2]	y[n-1]	y[n]	c	theta (degrees)			
3					0.85	55			
4	1	0	0	1					
5	0	0	1	0.9751					
6	0	1	0.9751	0.2283					
7	0	0.97508	0.2283	-0.4819					
8	0	0.22828	-0.4819	-0.6348					
9	0	-0.4819	-0.6348	-0.2708					
10	0	-0.6348	-0.2708	0.1946					
11	0	-0.2708	0.1946	0.3854					
12	0	0.19458	0.3854	0.2352					
13	0	0.38541	0.2352	-0.0491					
14	0	0.23522	-0.0491	-0.2178					
15	0	-0.0491	-0.2178	-0.1769					
16	0	-0.2178	-0.1769	-0.0151					
17	0	-0.1769	-0.0151	0.1131					
18	0	-0.0151	0.1131	0.1212					
19	0	0.11307	0.1212	0.0365					
20	0	0.12118	0.0365	-0.052					
21	0	0.03647	-0.052	-0.077					
22	0	-0.052	-0.077	-0.0376					
23	0	-0.077	-0.0376	0.019					
24	0	-0.0376	0.019	0.0457					

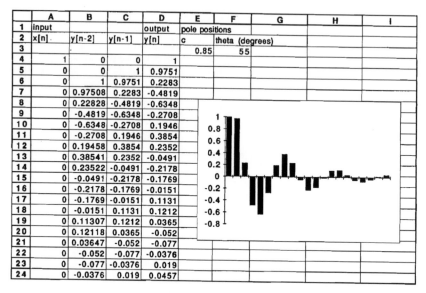

Fig. 12.7 Spreadsheet computation of unit sample response.

margin is about 44° at a crossover frequency of around $0.45\,\mathrm{rad\,s^{-1}}$, thus confirming the approximations of the manual graphing of Chapter 8. The parameters of the individual components (compensator time constants, plant natural frequency and damping ratio) are stored in row 2; changing the entries in the appropriate cells causes the complete spreadsheet to be recalculated. Built-in spreadsheet graphics functions are used to display the results automatically as a Nichols plot.

Spreadsheets are particularly useful for carrying out simple discrete system simulations, by automating the tabular procedure of Chapters 9 and 10. Figure 12.7 shows an example for computing the unit sample response corresponding to given complex pole positions. An iterative solution is computed to the second-order difference equation

$$y[n] = x[n] + 2c\cos\theta\, y[n-1] - c^2 y[n-2]$$

defined by an arbitrary pair of complex poles located at $z = c\exp(\pm j\theta)$. Here the spreadsheet automatically carries over the results of one line diagonally into the appropriate column of the next, as required by the technique of Chapter 9. Although such spreadsheets become unwieldy for higher-order systems, simple designs can be carried out in this way: even more important, perhaps, is the way spreadsheets can engender a deeper understanding by the user of the underlying models.

Non-linearities

Non-linear control systems are beyond the scope of this book, but it is worth noting that most dedicated control system design packages allow non-linear com-

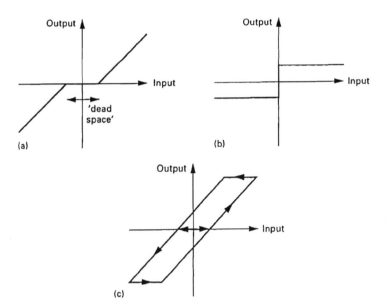

Fig. 12.8 Commonly encountered non-linearities: (a) dead space; (b) 'bang–bang' or switching element; (c) backlash.

ponents, such as those with the static input–output characteristics of Fig. 12.8, to be built into the system model.

Relays, with a characteristic like Fig. 12.8(b), and components exhibiting dead space (Fig. 12.8(a)) or backlash (Fig. 12.8(c)), are extremely common in control systems, and their presence can have a great effect on closed-loop behaviour: with the aid of computer-based tools the user can simulate such systems in a straight-forward way.

Generic mathematics packages and spreadsheets can also be used for such purposes. For specific applications of spreadsheets to non-linear system modelling, including a 'tutorial', see: Kolk, W.R. and Lerman, R.A., *Non-linear System Dynamics*, Van Nostrand Reinhold, 1992.

Computers in controllers

In the 1960s, digital computers began to be used in a supervisory mode to adjust the set points of the analogue (often pneumatic) controllers of individual process control feedback loops. If the mainframe computer failed, the individual loops functioned normally (with constant set points). Later, as digital technology became more reliable, and control engineers had more faith in computers, direct digital control (DDC) algorithms were introduced in which the computer itself generated the signals to be sent to the actuator, using algorithms of the type presented in Chapter 11. Today, both stand-alone digital controllers and industrial standard PCs have considerable computing power, and their control software can offer the user many facilities in addition to traditional three-term or phase compensation algorithms.

Adaptive control and self-tuners

As has been noted on a number of occasions in this book, plant models can only ever be approximate; indeed, the behaviour of the plant can change considerably with time. Various *adaptive control* techniques have been developed to cope with such situations. In one of the simplest, and earliest, adaptive techniques known as *gain scheduling*, different sets of controller parameters are stored in a look-up table and used in the appropriate circumstances. For example, the parameters used in an aircraft flight control system may be made to depend on altitude, since aircraft dynamics change with height. Similarly, the cooling effect of water inflow into a reactor is very different depending on whether 'flashing' (the sudden boiling of feedwater) takes place or not: gain scheduling can again be used effectively as part of the control system.

For a down-to-earth review of techniques see Smith, L.S.P.S., 'Adaptive control of industrial processes', *Trans. Inst. Measurement & Control*, Vol. 14, No. 5, 1992, pp. 238–242.

Gain scheduling depends on knowing in advance that particular, well-defined regimes exist, for which suitable controller parameters can be stored and implemented as required. Other adaptive control methods are designed to deal with plant models which vary less predictably. For example, adaptive robot arm controllers have been designed in which the behaviour of the arm under different load conditions is monitored using a system identification algorithm. Actual behaviour is then compared with that predicted by a stored model: if the stored model becomes inadequate, it can be modified automatically and controller parameters changed accordingly. Similarly, *self-tuning* process controllers are designed to adjust their own settings in response to changing plant conditions. They can do this in a variety of ways: one of the simplest is for the self-tuner to inject a small test step change into the plant when in self-tune mode, monitor the response, and automatically set its parameters according to Ziegler–Nichols rules or a similar algorithm.

Many of the classical compensator design and controller tuning techniques described in this book result in robust systems whose performance does not alter significantly as a result of slight changes to plant characteristics. Adaptive techniques can certainly produce higher-performance systems in some cases, but control engineers must be on their guard, ready to ask such questions as:

- Are there circumstances in which the adaptive controller can 'switch' unpredictably and undesirably between regimes? Can it result in a high-order control algorithm which generates excessive control action? Is it unnecessarily sensitive to modest changes in plant characteristics?
- Under what conditions is it appropriate to use the self-tune function of a particular controller? Should the controller already be 'near' appropriate settings? How does it converge to 'good' settings? What other aspects of plant operation (noise, non-linearities) are relevant?

Fuzzy controllers

If process control operators are observed in action, or questioned about how they control a plant manually, 'rules' such as the following often emerge:

> If the deviation is *small*, but increasing *quite quickly*, then the control action is initially increased *quite a lot* for a *short* period, and then reduced to a steady value.

Fuzzy logic is a way of representing the imprecise italicized terms in a mathematical form. In a *fuzzy controller*, control action is generated by applying rules like these, instead of using the more traditional mathematical algorithms described elsewhere in this text. Commercial fuzzy controllers are already available, and a number of applications have been reported in which fuzzy controllers cope successfully with extremely difficult control tasks. At the time of writing (1993) fuzzy control is often used in 'messy' situations in which no traditional mathematical model of the process is available – although many applications to systems with well-defined process models have also been implemented. Whether or not fuzzy control is superior to other approaches (such as adaptive or self-tuning PID control), and likely to replace them eventually in a wide range of applications, is still being debated.

General issues in computer control

The more complex computer-based control becomes, the more the control engineer needs to be aware of software engineering issues. Large-scale computer control involves *concurrency* (several programs running – or appearing to run – at the same time); *real-time operation* (actions have to be performed within specified times, in response to signals from the outside environment); and sophisticated alarms, safety routines, and *exception handling* (responses to software error conditions). As a result of these and other developments, a whole new sub-discipline of 'real-time control' has emerged.

For an accessible introduction to this subject area, see: Bennett, S., *Real-time Control Systems: an Introduction*, Prentice-Hall, 2nd Edition, 1993.

The validation of software for such purposes is an extremely difficult task – as is establishing exactly how different parts of the system should 'talk' to one another. A new generation of standards is emerging for communication between the various elements of distributed control systems. In such an *open system* it should be possible, say, for transducers from one manufacturer, actuators from another, and controllers from a third to be interconnected in a system running under the supervision of software from a fourth supplier. Developing and agreeing such standards is a very lengthy process, however, and at the time of writing many questions remain to be settled.

A final reminder: know your plant!

In addition to familiarity with a range of modelling and design techniques, and the ability to use appropriate computer tools sensibly, control engineers also need a thorough understanding of the plant they are attempting to control – and at least some capacity for lateral thinking! Quite often a satisfactory engineering solution to a control problem will depend on matters which have little to do with formal control theory. To conclude this book, and to emphasize for one last time the need to put its subject matter in context, the reader is invited to reflect for a moment on the following situations.

Scenario 1. To satisfy performance specifications a particular position control system needs to introduce an amount of phase lead which is beyond the capability of simple lead compensation. Various more sophisticated controller designs are contemplated and simulated using a computer-aided package. Then the engineer examines the mathematical models being used for individual system components. The actuator originally proposed has a fairly long time constant, and contributes significantly to the phase lag at the critical frequency. Using a slightly more expensive, but faster, actuator means that a simpler compensator easily fulfils the specification.

Scenario 2. A self-tuning controller sets PID parameters which successfully maintain the output of a process within specification, countering the effects of a disturbance. Unknown to the engineer, however, the disturbance arises from the faulty operation of a particular valve. Locating and remedying the fault would mean the same control system could achieve the same or better performance – but with much less control effort, less component wear, and a considerable saving on operating costs.

Scenario 3. A satellite consists of two bodies connected by a flexible link. The larger of the two bodies houses noisy power supply, thrusters, etc., while the smaller contains sensitive astronomical sensors which need to be isolated from the vibrations of the first. A preliminary design for an attitude control system has the transducer mounted with the astronomical equipment on the smaller body, and the actuator (thruster) on the larger. This proves to be a difficult design owing to lightly-damped mechanical resonances in the flexible structure. By moving the sensor to the same body as the actuator the control problem is reduced to a much easier one.

Franklin, Powell and Emami-Naeini give a detailed case study of such a system.

The lessons to be learnt from such examples include:

- It is sometimes possible to modify the plant instead of indulging in fancy controller design.
- It's the control engineer's business to know exactly what's going on – even when the system appears to be operating properly.
- Don't use a control system to compensate for a disturbance if you can remove the disturbance instead!
- Transducer location can be just as important as compensator design.

Summary

Computers have radically altered the way control systems are designed and implemented. Powerful software packages are available as an aid to control system design, and digital controllers offer facilities such as self-tuning in addition to PID and other control algorithms. For successful system design, a knowledge of control theory and modelling techniques needs to be supplemented by a thorough understanding of the practicalities of the plant to be controlled. Computer tools of various types can help with calculations and simulations, but it is the engineer who finally has to make the judgements – and ultimately, perhaps, defend them in a court of law!

Appendix 1
Polar plots

As an alternative to the Bode and Nichols plots described in Chapters 3 and 4, frequency response curves may be plotted using polar coordinates. Figure A1.1 illustrates the general idea: the amplitude ratio and phase shift for a given frequency ω_p are represented by the polar coordinates of a point P. The locus of all such points represents the complete frequency response. Note that a phase lag is represented by a clockwise rotation.

Note that the amplitude ratio is plotted as a numerical factor, not in decibels.

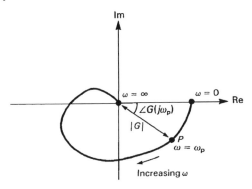

Fig. A1.1   Frequency response as a polar plot.

The phase shift of a transfer function model which is a ratio of polynomials will tend to a multiple of $-90°$ at high frequencies. Raw frequency response data, however, may differ considerably from this ideal; the polar plots given in this Appendix have therefore been sketched deliberately so as not always to conform to ideal models.

The simplified Nyquist criterion was originally stated with reference to such a polar plot. The critical point on the polar plot is the point $(-1, j0)$, corresponding to an amplitude ratio of 1 (0 dB) and a phase shift of $-180°$. The stability criterion can then be stated:

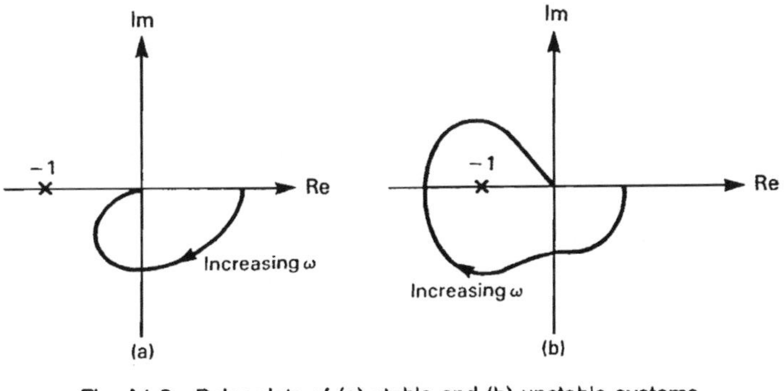

Fig. A1.2   Polar plots of (a) stable and (b) unstable systems.

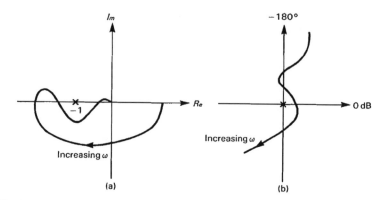

Fig. A1.3   A conditionally stable system (a) on a polar plot and (b) on a Nichols plot.

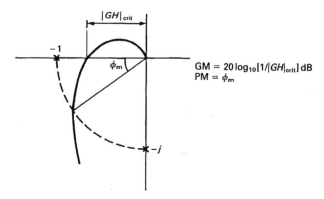

$$GM = 20 \log_{10}[1/|GH|_{crit}] \text{ dB}$$
$$PM = \phi_m$$

Fig. A1.4   Gain and phase margins.

If an open-loop system is stable, then the corresponding closed-loop system is also stable, provided that the open-loop frequency response curve does not enclose the $(-1, j0)$ point.

Figure A1.2 illustrates the meaning of the term 'enclose'. Figure (a) shows a stable system, whose frequency response locus does not enclose the critical point, while (b) represents an unstable system. Figure A1.3 shows a conditionally stable system, with a Nichols plot representation for comparison. On the polar plot, increasing the loop gain corresponds to expanding the curve by a given factor (cf. raising the curve on the Nichols chart); decreasing the gain corresponds to shrinking the polar plot (lowering the curve on the Nichols chart). Both representations therefore show that the system of Fig. A1.3 can become unstable if the gain is increased or decreased.

Figure A1.4 illustrates gain and phase margins on a polar plot. Increasing the phase lag corresponds to rotating the polar plot clockwise: the phase margin is therefore the angle of rotation necessary to cause the curve to pass through the critical point. Similarly the gain margin is the increase in gain necessary to 'expand' the curve sufficiently to pass through the $(-1, j0)$ point.

Appendix 2
The Routh–Hurwitz criterion

The Routh–Hurwitz criterion is used to decide, from the characteristic equation alone, whether or not a system is stable. The criterion determines simply whether all the roots of the characteristic equation have negative real parts (that is, whether all poles lie in the left half plane), and gives no other information on the nature of the roots or the relative stability of the system.

Suppose that the characteristic equation of a system is

$$a_0 s^n + a_1 s^{n-1} + \ldots a_{n-1} s + a_n = 0$$

In a closed-loop system this will be derived from the expression $1 + G(s) H(s) = 0$, where $G(s)$ and $H(s)$ are forward-path and feedback-path transfer functions.

Then the procedure for testing stability is as follows:

(1) Do all the coefficients of the characteristic equation have the same sign? If not, the system is unstable; if so, proceed to 2.
(2) Does the characteristic equation possess terms in all powers of s from 0 to n inclusive? If not, the system is unstable; if so, proceed to 3.
(3) Construct the Routh array as follows:

$$
\begin{array}{llll}
a_0 & a_2 & a_4 & a_6 \ldots \\
a_1 & a_3 & a_5 & a_7 \ldots \\
b_1 & b_2 & b_3 \ldots \\
c_1 & c_2 & c_3 \ldots \\
d_1 & d_2 \ldots
\end{array}
$$

where the first two rows are written down from the characteristic equation, and the third and subsequent rows are constructed according to the rules:

$$b_1 = a_2 - \frac{a_0}{a_1} \times a_3; \quad b_2 = a_4 - \frac{a_0}{a_1} \times a_5; \quad b_3 = a_6 - \frac{a_0}{a_1} \times a_7; \text{ etc.}$$

$$c_1 = a_3 - \frac{a_1}{b_1} \times b_2; \quad c_2 = a_5 - \frac{a_1}{b_1} \times b_3; \text{ etc.}$$

$$d_1 = b_2 - \frac{b_1}{c_1} \times c_2; \text{ etc.}$$

Cases sometimes arise where the presence of zeros in the early part of the array mean that the array cannot be completed. Special procedures exist for dealing with such cases; for details see the texts in 'Further Reading'.

and so on.

Continue calculating entries until only zeros are obtained, the rows shortening so that the final row contains only one value.

The system is stable if all the numbers in the left-hand column of the array are of the same sign and not equal to zero.

Example A2.1 The characteristic equation of a system is

$$s^4 + 2s^3 + 3s^2 + 4s + 5 = 0$$

254

(1) All coefficients have the same sign.
(2) All powers of s are present.
(3) Write down the first two lines of the array.

$$\begin{array}{ccc} a_0 & a_2 & a_4 \\ 1 & 3 & 5 \end{array}$$

$$\begin{array}{ccc} a_1 & a_3 & a_5 \\ 2 & 4 & 6 \end{array}$$

Construct row 3:

$$b_1 = a_2 - \frac{a_0}{a_1} \times a_3 = 3 - \frac{1}{2} \times 4 = 1$$

$$b_2 = a_4 - \frac{a_0}{a_1} \times a_5 = 5 - \frac{1}{2} \times 0 = 5$$

Since there is no explicit a_5, the value zero is used to calculate b_2.

$$b_3 = a_6 - \frac{a_0}{a_1} \times a_7 = 0$$

The array so far is therefore

$$\begin{array}{ccc} 1 & 3 & 5 \\ 2 & 4 & \\ 1 & 5 & \end{array}$$

Construct row 4:

$$c_1 = a_3 - \frac{a_1}{b_1} \times b_2 = 4 - \frac{2}{1} \times 5 = -6$$

There is no need to continue, since c_1 has a negative sign, indicating instability. For completeness, however, note that there is no c_2 and the final entry of the array, d_1, would be calculated as

$$d_1 = b_2 - \frac{b_1}{c_1} \times c_2 = 5 - \frac{1}{-6} \times 0 = 5$$

Example A2.2

The Routh–Hurwitz criterion can also be used to determine limiting values of system parameters such as gains or time constants, by constructing the array in partially algebraic form. For example, consider the system of Fig. A2.1. Then the characteristic equation is given by

$$1 + 0.05K/[s(s^2 - 3.4s - 0.062)] = 0$$

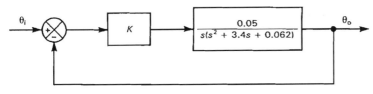

Fig. A2.1

or

$$s^3 + 3.4s^2 + 0.062s + 0.05K = 0$$

where K is the proportional gain.

Constructing the Routh array in terms of a general gain K we have:

$$
\begin{array}{ll}
1 & 0.062 \\
3.4 & 0.05K \\
0.062 - \dfrac{0.05K}{3.4} & \\
0.05K &
\end{array}
$$

This example is the ship autopilot of Chapter 8 under proportional control only. The limiting value of K obtained from the Routh–Hurwitz criterion agrees with the Nichols plot of Fig. 8.14, which indicated marginal instability for $K = 5$.

For stability, all the values in the left-hand column must be positive. Hence for positive K, the system is only stable so long as

$$K < (0.062 \times 3.4)/0.05 \approx 4.2$$

Appendix 3
Frequency response of discrete linear systems

The 'frequency preservation' property of a discrete linear system, which was stated in Chapter 11, is illustrated in Fig. A3.1. To determine the amplitude ratio B/A and the phase shift θ at any given frequency ω, the term $e^{j\omega T}$ is substituted for z in the transfer function $G(z)$ of the discrete system. The discrete amplitude ratio is then given by $|G|$ and the phase shift by $\arg G$.

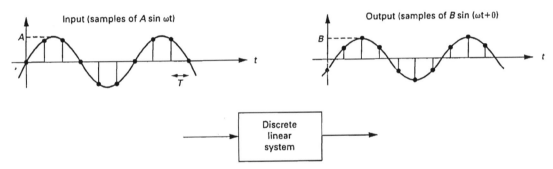

Input (samples of $A \sin \omega t$)

Output (samples of $B \sin (\omega t + \theta)$)

Discrete
linear
system

Fig. A3.1 'Frequency preservation' property of a discrete linear system.

Just as the amplitude ratio of a continuous linear system can be visualized as the 'cut' through the $|G(s)|$ surface along the $j\omega$ axis, so the amplitude ratio of a discrete linear system can be visualized as a suitable cut through the $|G(z)|$ surface.

> This approach was described in Chapter 5.

Figure A3.2 shows the pole–zero diagram and the $|G(z)|$ surface for the two-term averager of Chapter 9, with transfer function $G(z) = (z + 1)/2z$. The complete amplitude ratio of the averager is given by the value of $|G(z)||_{z \to e^{j\omega T}}$ as ω is increased indefinitely from zero. In order to visualize the value of $|G(z)|$ where $z = e^{j\omega T}$, consider first the points in the z-plane corresponding to these values. Viewing $e^{j\omega T}$ as the polar representation of a complex number, we see that its magnitude is always 1 and its angle, ωT, increases as ω increases. This is shown in Fig. A3.3: the values of $z = e^{j\omega T}$ all lie on the unit circle in the z-plane. The values of $|G(z)|$ representing the system amplitude ratio are therefore those corresponding to the height of the cut around this circle in the $|G(z)|$ surface. Such a cut is illustrated for the averager in Fig. A3.4 and for a simple recursive system in Fig. A3.5.

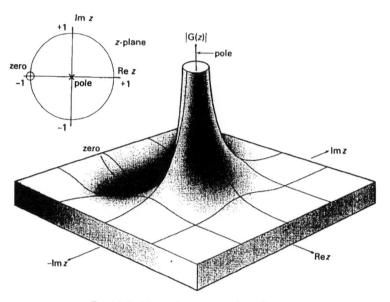

Fig. A3.2   The z-plane and the $|G(z)|$ surface.

The axes of Fig. A3.4 have been turned through 90° in comparison with Fig. A3.2, in order to show the region corresponding to $0 < \omega < \pi/T$ more clearly.

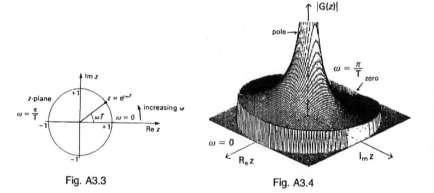

Fig. A3.3   Fig. A3.4

For more information on discrete frequency response see Meade and Dillon. For a non-mathematical introduction see Bissell, C.C., 'A phasor approach to digital filters', *Int. J. Elec. Eng. Ed.*, Vol. 24, pp. 197–213 (July 1987).

As can be deduced from Fig. A3.3, increasing ω from zero to half the sampling frequency $\omega = \pi/T$ is equivalent to tracing anti-clockwise half way round the unit circle, starting from the point $z = 1$. Increasing the frequency further brings us back to our starting point for a value of ω corresponding to the sampling frequency itself. Increasing ω *above* the sampling frequency corresponds to tracing round the cut a second time. Hence, as we saw in Chapter 9, the frequency response of the averager (or any discrete linear system) repeats at intervals of the sampling frequency. Note that in the region up to half the sampling frequency (the region free from potential aliasing problems) both the averager and the recursive system of Fig. A3.5 behave as low-pass filters.

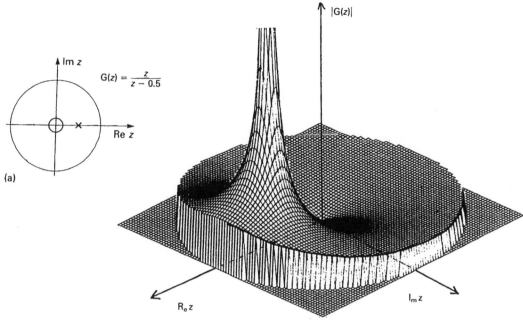

(a)

$G(z) = \dfrac{z}{z - 0.5}$

Fig. A3.5

Answers to numerical problems

2.1 (a) 0.84; (b) 1.04

2.2 $\dfrac{c}{r} = \dfrac{KG_1G_2}{1 + G_1(H + KG_2)}$

2.3 $p' = 300q'$

3.1 (a) $\tau = 1.25\,$s
   $k = 1.5$
   (b) $\tau = 0.2\,$s
   $k = 0.2$
   (c) $\omega_n = 2\,\text{rad s}^{-1}$
   $\zeta = 0.35$
   $k = 2$

3.4 (a) $\zeta = 0.2$
   $\omega_n = 98\,\text{rad s}^{-1}$
   (b) 50% approx.
   (c) 11 ms approx.

4.1 $GM \simeq 6\,$dB; $PM \simeq 17°$;
   closed-loop resonance peak $\simeq$
   10 dB

4.2 (b) 2.5; (c) 8 dB/50°; (d) 6.3;
   (e) zero

4.3 $K \simeq 2.2$ to 2.4; $T_i \simeq 0.8$ to 1.0 s

5.1 (a) $\dfrac{2s + 5}{(s + 2)(s + 3)}$; (b) $\dfrac{s + 2}{s(s + 1)}$;
   (c) $\dfrac{5}{s^2 + 25}$; (d) $\dfrac{s + 4}{s^2 + 8s + 20}$

5.2 (a) $6e^{-2t}$; (b) $0.5e^{-0.1t}$;
   (c) te^{-2t}; (d) $e^{-2t}\sin t$

5.3 (a) $2 - 3e^{-2t} + e^{-3t}$;
   (b) $1.4e^{-t}\sin(5t + \pi/4)$

5.4 (a) Poles at $s = -2, -0.3$;
   stable; $K_1e^{-2t} + K_2e^{-0.3t}$
   (b) Pole at $s = 1$, zero at $s = 0$;
   unstable
   (c) Poles at $s = -0.1, -0.2$;
   zero at $s = -1$; stable;
   $K_1e^{-0.1t} + K_2e^{-0.2t}$
   (d) Poles at $s = -0.75 \pm j3.1$;

$Re^{-0.75t}\sin(3.1t + \phi)$; stable
   (e) Poles at $s = \pm j2$; on stability
   borderline
   (f) Poles at $s = 0, -6$; on
   stability borderline

5.5 (a) both 2.5; (b) approximately 1

5.6 $\zeta \simeq 0.3$; $\omega_n \simeq 5\,\text{rad s}^{-1}$

5.7 (a) $\dfrac{\theta_o(s)}{D(s)} = \dfrac{0.1s}{s^2 + 0.6s + 0.125}$;
   disturbance step response
   $f(t) = 0.53e^{-0.3t}\sin 0.19t$
   $\dfrac{\theta_o(s)}{D(s)} = \dfrac{0.1s}{s^2 + 0.6s + 0.05}$;
   $f(t) = 0.25(e^{-0.1t} - e^{-0.5t})$

6.2 $s_g =$ (a) -0.225; (b) $+0.175$;
   $T_i < 1.54\,$s

6.3 $K = 24$

6.5 Additional term $(s + 15)$ in
   numerator of system of Fig. 6.32.
   The effect of the resulting closed-
   loop zero depends on its location
   with respect to closed-loop poles,
   and hence on K.

6.6 (a) $\dfrac{0.5}{0.3s^2 + (1 + 0.5K_v)s}$
   (b) $\dfrac{0.5K}{0.3s^2 + (1 + 0.5K_v)s + 0.5K}$
   (c) $K = 5.4$; $K_v = 0.52$

7.1 (a) zero; (b) 0.47 rad

8.1 $K \simeq 0.7$; $T_i = 3\,$s; $T_d = 0.75\,$s

8.3 $K = 20$; $C(s) = \dfrac{1 + 0.06s}{1 + 0.015s}$
   (or similar)

8.5 (a) $K = 15$;
   (b) $K = 13.5$,
   $T_i = 26.7$;
   (c) $K = 18$, $T_i = 16$, $T_d = 4$

260

9.1 $2 + 4z^{-1} + 6z^{-2} + 8z^{-3}$

9.2 $F(z) = 0.18z/(z - 1)(z - 0.82)$

9.3 $F(z) = 2z/(z - 20)$

9.4 Triple pole at origin; zeros at $z = -1$ and -0.25

9.5 $(z - 1)(z^2 + 4z + 1)/6Tz^3$

9.6 $y[n] = y[n - 1] + Te[n - 1]$; $G(z) = T/(z - 1)$

9.7 $0.74\,\text{Hz}$; $0.6\,\text{s}$

9.8 $y[m] = x[m] - 0.4x[m - 1] + 0.6y[m - 1]$

10.1 (a) $1.7/(z - 0.72)$; (b) $0.39/(z - 0.55)$

10.2 $y_{ss} = 0.89$; $e_{ss} = 0.11$

10.4 Step response samples are $0, 0.5, 0.75, 0.875, \ldots$

10.5 For $K = 0.5$, poles lie at $z = 0.5 \pm j0.5 \Rightarrow \zeta \approx 0.4$ and overshoot $\approx 25\%$. Simulated discrete step response: $0, 0, 0.5, 1, 1.25, 1.25, 1.125, \ldots$ Intersample behaviour

for an integrator plant obtained simply by joining samples with a straight line.

10.6 Zero at $z = -0.94$; poles at $z = 0.82, 1$

11.1 $\dfrac{0.32z^2 + 0.08z - 0.24}{z^3 - 1.68z^2 + 1.08z - 0.24}$

11.4 Peak overshoot $\approx 65\%$

11.5 (a) $(7z - 5)/(2.5z - 0.5)$;
(b) $0.2\,\text{rad}\ (\approx 11°)$;
(c) $(7.3z - 5.3)/(1.7z + 0.3)$ corresponding to $a = 9$ in the new continuous prototype compensator. (Slightly different numerical parameters may also be valid.) The 15% overshoot of (c) compared with >25% for (a) is possible only at the expense of greatly increased control action.

Further reading

Of the many standard texts covering control engineering and control theory, the following in particular approach the subject in a similar spirit to this Tutorial Guide.

1. Doebelin, E.O., *Control System Principles and Design* (Wiley, 1985).
2. Franklin, G.F., Powell, J.D. and Emami-Naeini, A., *Feedback Control of Dynamic Systems* (Addison-Wesley, 2nd ed. 1991).
3. Golten, J. and Verwer, A., *Control System Design and Simulation* (McGraw-Hill, 1991).
4. Leigh, J.R., *Control Theory. A Guided Tour* (Peter Peregrinus, 1992).

Those who require a more extensive knowledge of signal and systems theory are referred to:

5. Meade, M.L. and Dillon, C.R. *Signals and Systems* (Chapman & Hall, 2nd ed. 1991).

A thorough treatment of digital control, at a considerably more advanced level, can be found in:

6. Franklin, G.F., Powell, J.D. and Workman, M.L., *Digital Control of Dynamic Systems*, (Addison-Wesley, 2nd ed. 1990).

Index